WATER SUSTAINABILITY

A COMPREHENSIVE GUIDE FOR
EFFECTIVE WATER MANAGEMENT PRACTICES

I0396521

MRIDUL DEKA

INDIA • SINGAPORE • MALAYSIA

Notion Press Media Pvt Ltd

No. 50, Chettiyar Agaram Main Road,
Vanagaram, Chennai, Tamil Nadu – 600 095

First Published by Notion Press 2021
Copyright © Mridul Deka 2021
All Rights Reserved.

ISBN 978-1-63904-684-3

This book has been published with all efforts taken to make the material error-free after the consent of the author. However, the author and the publisher do not assume and hereby disclaim any liability to any party for any loss, damage, or disruption caused by errors or omissions, whether such errors or omissions result from negligence, accident, or any other cause.

While every effort has been made to avoid any mistake or omission, this publication is being sold on the condition and understanding that neither the author nor the publishers or printers would be liable in any manner to any person by reason of any mistake or omission in this publication or for any action taken or omitted to be taken or advice rendered or accepted on the basis of this work. For any defect in printing or binding the publishers will be liable only to replace the defective copy by another copy of this work then available.

CONTENTS

Preface 5

1 RESOURCES 7
    Water Resources throughout the regions of the world 9
    Status of water in South Asian countries 18
        Bangladesh 18
        Bhutan 18
        China 19
        Maldives 20
        Indonesia 21
        Mauritius 21
        Pakistan 22
        Myanmar 23
        Nepal 24
        Sri Lanka 24
        Indian Status 25
    Ground water status in India 27
    Rain The primary source 33
    Water pollution and effect on health 43
        Pesticides and Biocides 45

2 CRISIS 51
    Climate Change and Water Crisis 56
        Causes of Migrations from rural to urban areas 59
    Water Hot Spots and Geo Politics 61
        Jordan River Basin, Arab and Israel 61
        The Euphrates River basin, Turkey and Syria 62
        The Nile River Basin, Egypt and Ethiopia 62
        Israel and Palestine 64
        The Yellow River basin, Tibet, China and India 64
    Water scarce cities around the world 67
        Sao Paulo 67
        Chennai 68

    Jakarta 69
    Beijing 70
    Cairo 72
    Delhi 73
    Mexico City 74
    Cape Town 76
  Water conflict chronologies 77
  Water conflict involving Public Vs Private entity 80
    World Water Council 82
    Disappearing Aral Sea 83

3 **MANAGEMENT** 85
  Rain Water Harvesting Systems 87
    Roof Top Rain Water Harvesting Systems 87
    Components of Rain Water Harvesting Systems 88
  Traditional Rain Water Harvesting Systems 100
    Planning and Designing of Tanka 103
    Constructional features 104
    Major uses of Erys 111
  Traditional Water Harvesting Systems around the World 112
    Flood Water Harvesting 112
    Reducing water demand through Nitrogen fixing trees 115
    Reducing water demand through Waste Water Management 117
    Wastewater use for recharge of ground water 121
    Rate of Application of Wastewater for recharge 122
    Methods of Recharge 123
    Water reuse in Residential and Public Buildings 123
    Managing water demand through Evaporation and Seepage reduction . . 125
  Management of Water Losses in Water Supply Networks 133
    Contributors of water leakages in water mains 134
    Location of Leaks in pipe lines 136
    Infrastructure Leakage Index (ILI) 142
    Water loss management in large diameter pipes 143
  Case Studies in Water Loss Management 144
    Water Loss Management Efforts in Manila 147
    Managing apparent water losses through use of Unmeasured Flow
    Reducer (UFR) 149
  Strategic Management of Water Resources through Virtual Water Trade 150
    Virtual water trade from India 155
  Some good practices in water resources management in India 156
    Water Resources Management through Legislation 165
    Fundamental Human Right to Water 165
    Irrigation Laws 169
  Water legislation in countries 182

Bibliography *191*

PREFACE

This book has been written to sensitize the readers that actual utilizable water for living beings is a meagre percentage of the total quantity of water on earth. The available academic books on water being too technical in nature, elaborate on various water treatment methodologies or irrigation practices being followed. However, it is felt that existing books do not bring forth the impending crisis of water that looms large on face of the earth. This book tries to delve in to availability of water resources across the world and a future crisis we are looking at. The book takes stock of various water harvesting, ground water recharging methods and water saving techniques practiced around the globe without being too technical for easy grasp of the subject matter. The book gives an insight in to usefulness of water legislations with few legislations being in practice in India and abroad, it also highlights the strategic management of water resources though virtual water trade which may interest scholars, professionals and policy makers in dealing with management of water resources in future.

Mridul Deka
10.05.2021

Chapter 1

RESOURCES

"Man can live without clothes, without shelter, and some time without food; without water, however, he soon perishes'. It is not surprising, therefore that from the early periods of man's history, his dwelling places have been closely associated with lakes, rivers, springs and wells. Civilizations and cultures were nurtured in the valleys of the great rivers- the Nile, the Indus, the Ganges and the Yangtse"; Dr. Marcolino Gomes Candau, Director General, World Health Organization, International conference on 'Water for Peace', U.S.A., 1967.

All living species on earth need water to survive. They need water to regulate body temperature, to provide the means for nutrients to travel to organs and tissues. It also helps to transport oxygen to cells, removes wastes and protects joints and muscles. In short, water is the most essential constituent of all living species. Nearly fifty five percent of the body weight of an adult human being and up to ninety eight percent of the weight of some jelly fish is attributed to water. However, despite its visual abundance on the surface of earth, compared to the total volume, the water available for human use is minimal. It has now become scarce due to exponential growth of population over the years and over abstraction of existing water sources in many parts of the world.

Global Scenario: Three-fourth of the surface of earth is covered by water comprising an approximate volume of about 1.38 billion Km^3. However, 97.4 percent of all water i.e. about 1.34 billion Km^3 on earth lie in ocean or in form of sea or saline ground water. Only 2.6 percent of all water, i.e. about 36 million Km^3 on the planet are fresh water or considered usable. Of this again, nearly 24 million Km^3 are in form of Glaciers and snow caps. In the past three centuries, the withdrawal of fresh water has been almost 4.5 times the increase in population during the same period. Studies have shown that the global population has doubled since the year 1940, but the fresh water usage has increased four folds owing to various reasons, i.e. rapid urbanization, changes in living conditions, industrialization, etc. The primary source of all water on earth is the 108 thousand Km^3 of precipitation it receives annually. The total annual renewable fresh water supply, blue water flows on earth are estimated at 39.60 thousand Km^3. Of this total renewable water supply on earth due hydrological cycle, only 29.7 thousand Km^3 are accessible to human beings. (Blue water is the source of supply. It is equivalent to the natural water resources i.e. surface water and ground water runoff).

State of world's water:

Water Source	Water volume in cubic miles	Water volume in cubic Kms	Percent of Fresh Water	Percent of Total water
Oceans, Seas, & Bays	321,000,000	1,338,000,000	—	96.54
Ice caps, Glaciers, & Permanent Snow	5,773,000	24,060,000	68.6	1.74
Groundwater	5,614,000	23,400,000	—	1.69
Fresh	2,526,000	10,530,000	30.1	0.76
Saline	3,088,000	12,870,000	—	0.93
Soil Moisture	3,959	16,500	0.05	0.001
Ground Ice & Permafrost	71,970	300,000	0.86	0.022
Lakes	42,320	176,400	—	0.013
Fresh	21,830	91,000	0.26	0.007
Saline	20,490	85,400	—	0.007
Atmosphere	3,095	12,900	0.04	0.001
Swamp Water	2,752	11,470	0.03	0.0008
Rivers	509	2,120	0.006	0.0002
Biological Water	269	1,120	0.003	0.0001

Source: Peter H. Gleick (editor), 1993, Water in Crisis: A Guide to the World's Fresh Water Resources. (Oxford University Press, New York)

Data available from FAO, AQUASTAT online suggests that Oceania continent (comprising Australia New Zealand and Pacific Islands) has only 2.1 percent of world's internal renewable fresh water resources but the region has the highest per capita availability at 33469m³. Americas (North, Central & South including Brazil) has the highest share of world's internal renewable freshwater resources at 45.4 percent. Other continents with relevant percentage of world's fresh water sources are-

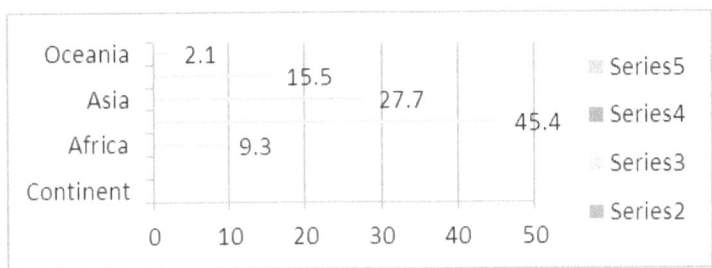

Per capita internal renewable fresh water resources (in m³) of all continents as in the year 2008 are as under-

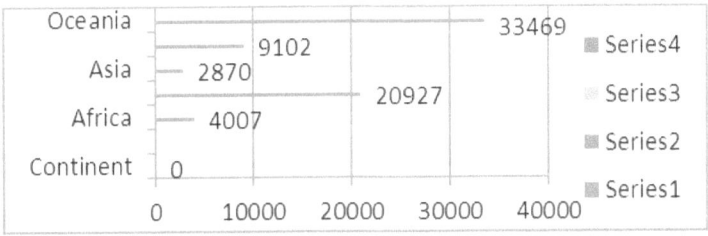

Source: http://www.fao.org/nr/aquastat

India, Bangladesh, Brazil, Canada, China, Colombia, Republic of Congo, Indonesia, Myanmar, USA, and former USSR are the countries whose annual renewable fresh water supply exceeds one thousand Km³.

WATER RESOURCES THROUGHOUT THE REGIONS OF THE WORLD

Canada: Canada has a total area of 9,976,180 Km². The fresh water bodies cover 7.6 percent of its areas. The precipitation is higher in Pacific coast and lower in the interior and Arctic and Sub Arctic regions. The precipitation ranges from 125mm in Arctic islands to more than 1482mm in Newfoundland. Snow contributes up to 23 percent of total precipitation in Montreal. Canada has approximately 9 percent of world's fresh water resources. The Great Lakes contain an estimated 22,700 km^3 of fresh water, 99 percent of which is non-renewable. The country has more than 31,000 fresh water lakes varying from 3 km^2 to 100km^2 in sizes.

United States of America: The United States of America comprises of 50 states covering an area of 9,363,368 Km^2 including Alaska and Hawaii. 48 states (excluding Alaska and Hawaii) forming a belt of conterminous states covering an area of 7,828,016 Km^2 extends from Atlantic Ocean to Pacific Ocean. These conterminous states receive precipitation ranging, from 450 mm in the Pacific coast to 710 mm in the Rocky Mountains. The region receives an average of 762 mm of rainfall annually. The USA is considered a water rich country, as the region representing 16 percent of world's total land surface, receives 17 percent of total precipitation on earth. The country is home to 7 percent of world's population and has a per capita fresh water of 16,000 m^3/year, which far exceeds the world average.

Caribbean Islands: The Caribbean Islands are categorized in to two groups; the Greater Antilles and the Lesser Antilles. The Greater Antilles comprises of Haiti, Cuba, Jamaica and the Dominican Republic. The Greater Antilles covers an area of 198,330 Km^2 and has a population of 30 million in 2000. Over 60 percent of the population live in urban areas. Cuba is the largest island with a population density of 101 inhabitants per Km^2.

Lesser Antilles comprises of Antigua and Barbuda, Barbados, Dominica, Grenada, Saint Kitts and Nevis, Saint Lucia, Saint Vincent and the Grenadines, and Trinidad and Tobago. The Lesser Antilles has a total area of 8430Km^2 receiving an average annual rainfall of 1141mm.

Caribbean Islands region is well endowed with water resources. The region receives 1.4 percent of world's precipitation and generates 1.8 of its water resources. With a population of 1.1 percent of world's population, the Caribbean Islands has about 11,900 m^3/year per capita fresh water resources.

Brazil - South America: The total area of the South American region is about 17.80 million Km^2, of which 65 percent area is covered by Argentina, Brazil and Mexico. Brazil has a total area of 8.547 million Km^2. The south of Brazil has a uniform climate and rainfall varies from 1250 – 2000 mm per year. In the northeast, the rainfall ranges from 900 to 4400 mm/year. The north east includes semi-arid lands of Brazil with irregularly distributed rainfall ranging from 250 mm to 750 mm annually. The north of Brazil covers almost whole of Amazon River basin. The climate in the basin is hot and humid with average temperature of 24° -26° C and a rainfall of 1500mm to 3000 mm per year.

The entire South American region is home to world's 5.7 percent population and it receives 26 percent of world's precipitation and the per capita fresh water resources are estimated at 35,000 m^3/year.

Western and Central Europe: The Western and Central Europe have 8.4 percent of world population and a land area of 3.7 percent of the world's total land area. The Western and Central Europe region include all of Europe with exception of Turkey and countries of former Soviet Union. The region comprises of Denmark, Finland, Norway, Sweden, Austria, Belgium, France, Germany, Ireland, Netherlands, Switzerland, United Kingdom, Bosnia and Herzegovina, Bulgaria, Croatia, Czech Republic, Hungary, Poland, Romania, Greece, Italy, Portugal and Spain.

The region has 5 percent of world's water resources at approximately 2200 Km^3. The per capita availability is 4342 m^3/year. The distribution of precipitation in the region is very diverse, ranging from less than 300 mm/year in many Mediterranean plains to more than 3000 mm/year on the coast of Norwegian Sea or in some areas of Balkans. The wetter regions are found along the Atlantic shores from Spain to Norway, the Alps and their eastern extension. There are seven major water basins each with an area of 100,000 Km^2 or more. The Danube River basin covering 17 percent of the region's total area with 800,000 Km^2 is the largest in the region. Rhine River basin covering an area of 5 percent of the area is the second largest. Most of the rivers are shared by two or more nations. The Danube River basin is shared by 13 countries, whereas Rhine River basin is shared by 7 countries. Croatia, Hungary, Luxembourg, the Netherlands, Portugal, Romania, Slovakia, Slovenia,

Yugoslavia are some countries, those depend on their neighbours for water. There is uneven distribution of water resources in the region due to uneven distribution of population and high urban growth in some countries. In 2000, per capita renewable water resources varied from 40m³/year in Malta to 992 m³/year in Cyprus, from 1140 m³/year in Denmark to 85,000 m³/year in Norway and more than 600,000 m³/year in Iceland. Considering that water resources remain unexploited in many parts of the region, the per capita availability of 4342 m³/year is not truly representative considering its high diversity. Some countries relying heavily on external water resources would fall below 1000 m³/year/capita limit, if they have to depend on internal water resources like Hungary (with less than 600 m³/year/capita) and the Netherlands (with less than 700 m³/year/capita).

Classification of Western and Central Europe countries by water resources:

Resources per capita (m³/year)	
Low- (< 1000m³/year)	Cyprus, Malta.
Average- (1000-10,000m³/year)	Denmark, France, Italy, United Kingdom, Germany, Greece, Netherlands, Portugal, Austria, Bulgaria, Luxembourg, Romania, Bosnia and Herzegovina, Slovakia, Spain, Switzerland.
High- (10000-100,000m³/year)	Finland, Ireland, Norway, Sweden, Albania, Croatia, Hungary, Slovenia, Yugoslavia.
Very High-(>100,000m³/year)	Icelnad.

Source: FAO Corporate Document Repository.

Africa: The continent of Africa comprises of 53 countries covering 22.4 percent of world's land area. The continent has 8 major river basins draining in to sea – Senegal, Niger, Nile, Shebelle-Juba, Congo, Zambezi, Limpopo and Orange Rivers. Africa has 9 percent of world's water. As a whole, the continent has 4979 m³/year of per capita water resources with high diversity across the region where Northern Africa and Sudano-Sahellian sub region contributes to only 1.2 percent and 4.06 percent respectively of the total water resources of the continent. The Congo River basin alone contributes 30 percent of the continent's run-off. In Sudan, Mali and Botswana, the quantity of water flowing in to the country exceeds the quantity leaving it. The evaporation rate being high, the river flows in these countries diminishes as they flow downstream. The Nile, Niger, Chari, Senegal, Okavango Rivers all carry water from wet to dry regions. The countries of the Sahel in western Africa receive approximately 50Km^3/year of water from the neighbouring countries of the south. The Saharan countries such as Algeria, Egypt, Libyan Arab Jamahiriya, Niger and Tunisia, has large non-renewable ground water resources which can be of sevral billion cubic meters. They are located in large sedimentary aquifer systems like in Nubian Sandstones, Sahel, Chad watersheds, Kalahari etc. The Africa region can be divided in to seven sub regions based on geography and prevailing climate:

Northern Africa: Comprising of Algeria, Egypt, Arab Jamahiriya, Morocco and Tunisia.

This region falls in the situation of very severe water scarcity with very limited water resources, receiving precipitation of less than 10mm/year. The fresh water resources per inhabitant, ranges between 200m^3/year to 700m^3/year. It is the poorest sub region in terms of internal water resources and contributes only 1.2 percent of the continent's total water resources. It is also the sub region with highest external water resources (63 percent) primarily due to Nile River. The sub region has large fossil water reserves in Continental Sahara, Murzuk, and Nubian Basin.

Sudano-Sahellian: Comprising of Burkina Faso, Cape Verde, Chad, Djibouti, Eritrea, Gambia, Mali, Mauritania, Niger, Senegal, Somalia, and Sudan.

The water resources of this sub region is limited, with 57 percent of its resources originating outside this sub region mostly in Fouta Djalon in Guinea (source of Gambia, Senegal, Niger, and Chari River) and in highlands of eastern Africa (Nile, Shebelle, and Juba Rivers). The sub region has less than 5 percent of the continent's internal water resources. This region also has fossil water like in Northern Africa, in Senegal-Mauritania Basin, the Lullemeden Basin in Niger, and the Chad Basin.

Gulf of Guinea: Comprising of Benin, Cote d'Ivoire, Ghana, Guinea, Guinea-Bissau, Liberia, Nigeria, Sierra Leone, and Togo.

This sub region has monsoon type, tropical humid climate. The main river basin is the Niger, which is shared by 10 countries. The sub region's internal water resources contribute 25 percent of the continent's total water resources. Ground water represents 33 percent to 50 percent of the sub region's total water resources.

Central Africa: Comprising of Angola, Cameroon, Central African Republic, Congo, Democratic Republic of Congo, Equatorial Guinea, Gabon, Sao Tome, and Principe.

Central African region has humid equatorial climate in the south with long rainy season. Two major rivers, the Congo River and the Ogooue River run through this sub region. This sub region is considered water rich having 48.4 percent of Africa's total water resources. The sub region caters to water needs of neighbouring sub regions.

Eastern Africa: Comprising of Burundi, Ethiopia, Kenya, Rwanda, Tanzania, and Uganda.

The water resources in Eastern Africa sub region are scarce. It has only 6.5 percent of the continent's total water resources. There are no major rivers other than the Nile in Ethiopia. The sub region does not receive much water from outside. However, it has the continent's largest lake: Lake Victoria. The Eastern African sub region provides water to Sudano- Shellian sub region through the Nile River.

Indian Ocean Islands: Comprising of Comoros, Madagascar, Mauritius, and Seychelles.

This sub region containing Madagascar and other small island nations ranks third in terms of water resources. Madagascar ranks second in terms of water resources in the continent.

Southern Africa: Comprising of Botswana, Lesotho, Malawi, Mozambique, Namibia, South Africa, Swaziland, Zambia, and Zimbabwe.

This sub region has very diverse climate from sub-tropical humid to arid. It contributes to 7 percent of continent's total water resources. External water resources enter from Central Africa through Zambezi basin.

Near East: Based on their common characteristics, the Near East region can be divided in to three sub regions-

Arabian Peninsula: Comprising of Bahrain, Kuwait, Oman, Qatar, Saudi Arabia, United Arab Emirates, and Yemen.

Caucasus: Comprising of Armenia, Azerbaijan, and Georgia.

Middle East: Comprising of Gaza Strip, Iraq, Islamic Republic of Iran, Israel, Jordan, Lebanon, Syrian Arab Republic, Turkey, and West Bank.

The Near East region has a population of about 250 million inhabitants which is 4.25 percent of total population on earth and covers an area of 6.34 million Km^2, which is approximately 4.70 percent of world's total. With mere 1.1 percent of total renewable water resources on earth and a comparatively higher population, it is one of the most water scarce regions on the planet, in terms of per capita availability. In 10 out of 18 countries in the region, the fresh water availability is less than 1000m^3/inhabitant/year. The Arabian Peninsula receives less than 10mm/year of rainfall and the water availability is 200m^3 per inhabitant per year to 700m^3/inhabitant/year. The three sub regions of the Near East region are independent of one another in terms of their water resources, with no exchange of water resources between them. Due to their favourable geographical situation, the Islamic Republic of Iran and Lebanon has internal renewable water resources more than 1000m^3/inhabitant/year. These two countries along with Turkey contribute heavily to water resources of the downstream countries through river systems.

Iraq and Syrian Arab Republic depends more than 50 percent of their water recourses to external sources. Israel and Jordan have water resources less than 1000m^3/inhabitant/year. These countries face acute shortage of water resources.

Central Asia: The Central Asian region consists of Aral Sea countries of Kazakhstan and Uzbekistan and other countries which comprise Afghanistan, Kyrgyzstan, Tajikistan, and Turkmenistan. The region has 1.3 percent of world's population and 3.5 percent of land area. The water resources available per inhabitant are 3320m^3/year. Its total renewable water resources are only about 0.7 percent of the world's total. The Aral Sea is located in a depression of Turan Plain and fed by two major rivers: Amu Darya in the south and Syr Darya in the north. These rivers originate in Pamir and Tien Shan mountain ranges respectively. The combined basin area of these rivers is 1.9 million Km^2. The flow in these rivers reaching Aral Sea during 1960 was estimated at 45-50Km^3/year. However due to large scale exploitation for irrigation, the flow has reduced to a present level of around 6-12Km^3/year. The population of Aral Sea basin rose from 14 million in 1960 to 27 million in 1980; with substantial increase in irrigation area from 4.7 million in 1960 to 7 million in 1980.

Southern and Eastern Asia: The Southern and Eastern Asian region has 15 percent of the world's total land area which covers about 20.4 million Km^2. The Indian subcontinent region comprises of Bangladesh, India, Bhutan, Maldives, Nepal, Pakistan, and Sri Lanka. The sub region extends over an area of 3,961, 680 Km^2. This sub region consists of the large portion of flood plains along Indus and Ganges river basins. The sub region experiences tropical monsoon climate, with significant variance in rainfall and temperature. About 80 percent of rainfall occurs during summer monsoon season (June-September). The average annual precipitation in the sub region is about 1279mm, which varies from 150mm in North West desert of Rajasthan, India to more than 10,000mm in the Khasi Hills in North East India.

The Eastern Asia sub region consists of China, Mongolia, and the Democratic People's Republic of Korea. It covers an area of 11,285,070 Km^2. Most of this sub region is mountainous with about 80 percent lying 1000mtrs above sea level. The monsoon determines the climate in eastern and southern China. In rest of the sub region, climate is characterized by long cold winter attributed mainly to northern and north western winds from Siberia. Large part of southern Mongolia and central China are influenced by very arid climate and experience water scarce conditions. The rainfall ranges from scanty 25mm in the Tarim and Qaidam basins in China to 1520mm in Democratic People's Republic of Korea. The sub region has an average annual rainfall of 597mm.

Far East: The Far East sub region includes Japan and Republic of Korea. The sub region has a total area of 477,060 Km^2. 70 percent of the sub region are mountainous. The mountainous topography divides the land mass in to two distinct climatic zones; the Pacific Central zone characterized by summer monsoon blowing from Pacific Ocean and bringing rain along, and the Continental zone characterized by winter monsoon from Asian continent which bring heavy snowfall. Most precipitation occurs during June to September with an average annual precipitation of 1634mm.

South East Asia: The Southeast Asia sub region is comprised of Myanmar, Cambodia, Thailand, Vietnam and Lao PDR. The sub region has an area of 1,939,230 Km^2. About two-third of this sub region is covered by mountains and hills. The region is influenced by south-west monsoon during May-October which is the wet season period and receives heavy rainfall. The dry season period is between November-February, which is characterized by north-east monsoon. The water level in the Mekong River may differ by up to 20 meters between dry and wet seasons. The average annual rainfall in the sub region is 1877mm, which ranges from 500mm in central dry zone of Myanmar to 4000mm in Bac Quang in Viet Nam.

Oceania and Pacific: The Oceania and Pacific region cover 6 percent of the world's total land area, at about 8 million Km^2 and 0.4 percent of the world's population. The region comprises of Australia, Fiji, Hawaii, New Caledonia, New Zealand, Polynesia, Solomon Islands, and the Pacific Islands. Being geographically separated, the islands have no exchange of water resources between themselves. Australia has one of the largest aquifer systems in the world, the Great Artesian Basin (GAB) having an estimated area of 1.7 million Km^2 and a storage volume of about 8.7 Km^3 of fresh water.

STATUS OF WATER IN SOUTH ASIAN COUNTRIES
(Source: FAO AQUASTAT)

Bangladesh
Precipitation	: 2666mm/year
Area of Country	: 14400x1000ha.
Precipitation	: 383.9Km^3/year
Total Internal Renewable Water Resources (TIRWR) :	
Surface water produced internally	: 83.91BCM (a)
Ground water produced internally	: 21.09BCM (b)
Overlap between surface and ground water	: 0 (c)
Total Internal Renewable Water Resources (TIRWR)	: 83.91 + 21.09 – 0
	: 105BCM (d)
External Renewable Water Resources (ERWR):	
Surface water entering and bordering	: 1122BCM (e)
(From India Ganges=525.02BCM,	
India Brahmaputra=537.24BCM,	
India Meghna/Barak=48.36BCM,	
Other outside GBM to Chittagong	: 11BCM
Ground water entering the country	: 0.032BCM (f)
Total Internal Renewable Water Resources (TIRWR)	: 1122BCM
Total Renewable Water Resources (TRWR)	
Surface water	: 1122 + 83.9
	: 1206BCM
Ground Water	: 21.09 + 0.032
	: 21.12BCM
Overlap between surface and ground water	: 0
Total Renewable Water Resources (TRWR	: 1206 + 21.12 – 0
	: 1227BCM
Dependency Ratio	: (e + f)/ (e + f + d) x 100
	: 91.44%

Bhutan
Precipitation	: 2200mm/year
Area of Country	: 3839 x 1000ha
Precipitation	: 84.46Km^3/year

Total Internal Renewable Water Resources:
Surface water produced internally : 78BCM (a)
Ground water produced internally : 7.8BCM (b)
Overlap between surface and ground water
 : 7.8BCM (c)

Total Internal Renewable Water Resources
(TIRWR) : 78BCM (d)
 (= a + b - c)

External Renewable Water Resources (ERWR):
Surface water entering the country : 0
Surface water entering and bordering : 0 (e)
Surface water leaving the country : 78BCM
Ground water entering the country : 0 (f)
Total Renewable Water Resources (TRWR)
Surface water : 78BCM
Ground water : 7.8BCM
Overlap between ground and surface water : 7.8BCM
Total Renewable Water Resources (TRWR) : 78 + 7.8 – 7.8
 : 78BCM

Dependency Ratio: (e + f)/ (e + f + d) : 0+0/0+0+78 = 0%

China
Internal Renewable Water Resources:
Precipitation : 645mm/year
Area of the country : 960000x1000ha
Precipitation : 6192Km^3/year
Surface water produced internally : 2712BCM (a)
Ground water produced internally : 828.8BCM (b)
Overlap between surface and ground water : 727.9BCM (c)
Total Internal Renewable Water Resources : 2813BCM (d)
 (= a+b-c)

External Renewable Water Resources (ERWR)
Surface water entering country : 17.17BCM
(From -
India =0.117, Kazakhstan = 2.327,
Kyrgyzstan=5.356,
Mongolia=1.401, Pakistan=0.718,
Vietnam=7.250)

Accounted inflow from Border Rivers	: 10.15BCM
Surface water entering and bordering	: 17.17 + 10.15 = 27.32BCM (e)
Surface water leaving the country (Outflow to-India Indus=181.62, India Brahmaputra=165.4BCM, Kazakhstan=21.54BCM, Kyrgyzstan=0.558, Lao & Myanmar=73.63, Myanmar=100.03, Nepal Ganges=12, Russian Federation=119.04, Vietnam=44.1)	: 717.9BCM
Total External Renewable Surface Water	: 27.32
Ground water entering the country	: 0 (f)
Total Renewable Water Resources (TRWR) Surface water	: 2713 + 27.32 = 2740BCM
Ground water	: 828.8BCM
Overlap between Surface and Ground water	: 727.9BCM
Total Renewable Water Resources (TRWR)	: 2740 + 828.8 − 727.9 = 2841BCM
Dependency Ratio: (e + f)/(e + f + d)	: (27.32 + 0)/(27.32 + 0 + 2813) = 0.9%

Maldives

Internal Renewable Water Resources (IRWR)

Precipitation	: 1972mm/year
Area of country	: 30 x 1000ha
Precipitation	: 0.59Km^3/year
Surface water produced internally	: 0 (a)
Ground water produced internally	: 0.03 (b)
Overlap between surface and ground water	: 0 (c)
Total Internal Renewable Water Resources	: 0.03 (d) (=a+b-c)

External Renewable Water Resources

Surface water entering the country	: 0
Surface water entering and bordering	: 0 (e)
Surface water leaving the country	: 0
Ground water entering the country	: 0 (f)

Total External Renewable Water Resources : 0
Total Renewable Water Resources (TRWR)
Surface water : 0
Ground water : 0.03BCM
Overlap between ground and surface water : 0
Total Renewable Water Resources : 0 + 0.03 − 0
= 0.03BCM

Dependency Ratio: (e + f)/ (e + f + d) : (0 + 0)/ (0 + 0 + 0.03) = 0%

Indonesia
Internal Renewable Water Resources:
Precipitation : 2702mm/year
Area of the country : 190457x1000ha.
Precipitation : 5146Km^3/year
Surface water produced internally : 1973BCM (a)
Ground water produced internally : 457.4BCM (b)
Overlap between surface water and ground water : 411.7BCM (c)
Total Internal Renewable Water Resources : 2019BCM (d) (=a+b-c)

External Renewable Water Resources:
Surface water entering the country : 0
Surface water entering and bordering : 0 (e)
Ground water entering the country : 0 (f)
Total Renewable Water Resource:
Surface water : 1973BCM
Ground water : 457.4BCM
Overlap between surface water and ground water : 411.7BCM
Total Renewable Water Resources :
(1973 + 457.4 - 411.7BCM)
: 2019BCM

Dependency Ratio: (e+f)/(e+f+d) : 0%

Mauritius
Internal Renewable Water Resources:
Precipitation : 2014mm/year
Area of the country : 204x1000ha.
Precipitation : 4.164Km^3/year
Surface water produced internally : 2.358BCM (a)

Ground water produced internally	: 0.893BCM (b)
Overlap between surface water and ground water	: 0.5BCM (c)
Total Internal Renewable Water Resources	: 2.751BCM
	(d) (=a+b-c)

External Renewable Water Resources:

Surface water entering the country	: 0
Surface water entering and bordering	: 0 (e)
Ground water entering the country	: 0 (f)

Total Renewable Water Resource:

Surface water	: 2.358BCM
Ground water	: 0.893BCM
Overlap between surface water and ground water	: 0.5BCM
Total Renewable Water Resources	: (2.358 + 0.893 - 0.5BCM)
	: 2.751BCM
Dependency Ratio: (e+f)/(e+f+d)	: 0%

Pakistan

Internal Renewable Water Resources:

Precipitation	: 494mm/year
Area of the country	: 79610x1000ha.
Precipitation	: 393.3Km^3/year
Surface water produced internally	: 47.4BCM (a)
Ground water produced internally	: 55BCM (b)
Overlap between surface water and ground water	
	: 47.4BCM (c)
Total Internal Renewable Water Resources	:55BCM
	(d) (=a+b-c)

External Renewable Water Resources:

Surface water entering the country	: 265.1BCM
(From Afghanistan=21.5; From India=243.58BCM)	
Inflow not submitted to treaties	: 21.5BCM
Inflow submitted to treaties	: 243.58BCM
Inflow secured through treaties	: 170.3BCM
(From western tributaries of Indus coming in From India)	
Surface water entering and bordering	: 191.8 (e)
Ground water entering the country	: 0 (f)

Total Renewable Water Resource:

Surface water	: 47.4 + 191.8

	: 239.2BCM
Ground water	: 55BCM
Overlap between surface water and ground water	: 47.4BCM
Total Renewable Water Resources	: 239.2 + 55 − 47.4
	: 246.8BCM
Dependency Ratio: (e+f)/(e+f+d)	: (191.8 + 0)/(191.8 +o+55)
	: 77.71%

Myanmar

Internal Renewable Water Resources:

Precipitation	: 2091mm/year
Area of the country	: 67659x1000ha.
Precipitation	: 1415Km^3/year
Surface water produced internally	: 992.1BCM (a)
Ground water produced internally	: 453.7BCM (b)
Overlap between surface water and ground water	: 443BCM (c)
Total Internal Renewable Water Resources	: 1003BCM
	(d) (=a+b−c)

External Renewable Water Resources:

Surface water entering the country	: 128.2BCM
(Inflow from India=20; Chine,Nu to Salween=68.74, Rivers in west Yunnan=31.29; Thailand=8.156)	
Flow in Border Rivers	: 73.63BCM
Accounted inflow in Border Rivers	: 36.81BCM
(Considering 50% of total flow)	
Surface water entering and bordering	: 128.2 + 36.81
	: 165BCM (e)
Ground water entering the country	: 0 (f)
Total External Renewable Water Resources	: 165BCM
Total Renewable Water Resource:	
Surface water	: 992.1 + 165
	: 1157.1BCM
Ground water	: 453.7BCM
Overlap between surface water and ground water	
	: 443BCM
Total Renewable Water Resources	: (1157.1 + 453.7 − 443)
	: 1168BCM
Dependency Ratio: (e+f)/(e+f+d)	(165+0)/(165+0+1003)
	: 14.13%

Nepal

Internal Renewable Water Resources:
Precipitation	: 1500mm/year
Area of the country	: 14718x1000ha.
Precipitation	: 220.8Km^3/year
Surface water produced internally	: 198.2BCM (a)
Ground water produced internally	: 20BCM (b)
Overlap between surface water and ground water	: 20BCM (c)
Total Internal Renewable Water Resources	: 198.2BCM (d) (=a+b-c)

External Renewable Water Resources:
Surface water entering the country (From China=12BCM)	: 12BCM
Surface water entering and bordering	: 12BCM (e)
Ground water entering the country	: 0 (f)

Total Renewable Water Resource:
Surface water	: 198.2 + 12
	: 210.2BCM
Ground water	: 20BCM
Overlap between surface water and ground water	: 20BCM
Total Renewable Water Resources	: 210.2 + 20 - 20
	: 210.2BCM
Dependency Ratio: (e+f)/(e+f+d)	: (12+0)/(12+0+198.2)
	: 5.71%

Sri Lanka

Internal Renewable Water Resources:
Precipitation	: 1712mm/year
Area of the country	: 6561x1000ha.
Precipitation	: 112.3Km^3/year
Surface water produced internally	: 52BCM (a)
Ground water produced internally	: 7.8BCM (b)
Overlap between surface water and ground water	: 7BCM (c)
Total Internal Renewable Water Resources	: 52.8BCM (d) (=a+b-c)

External Renewable Water Resources:
Surface water entering the country	: 0

Surface water entering and bordering	: 0 (e)
Ground water entering the country	: 0 (f)

Total Renewable Water Resource:

Surface water	: 52 + 0
	52BCM
Ground water	: 7.8BCM
Overlap between surface water and ground water	
	: 7BCM
Total Renewable Water Resources	: 52 + 7.8 − 7
	: 52.8BCM
Dependency Ratio	: (e+f)/(e+f+d):
(0+0)/(0+0+52.8)	: 0%

Indian Status

The status of total renewable fresh water resources in India is as under:

Surface water	: 1404.0BCM (a)
(Produced internally)	
Surface water entering country	: 635.20BCM
(From Nepal: 210.20BCM,	
Bhutan: 78.00BCM,	
China: 181.62BCM (Indus),	
165.40BCM (Brahmaputra))	
Surface water leaving the country	
Western tributaries of Indus	: 243.58BCM
Less	: 73.31BCM
(Reserved for India)	
Net	: 170.30BCM
Net Surface water entering country	: 635.20BCM
Less	: 170.30BCM
	: 464.90BCM
Total Renewable Surface Water	: 1869.0BCM
(1404.00BCM + 464.90BCM)	
Ground Water	: 432.20BCM (b)
(Produced internally)	
Overlap between Surface and Ground Water	: 390.00BCM (c)
Total Internal Renewable	
Water Resources (d)	: 1911.0BCM
(a+b-c) = (1404.04 + 432 − 390) BCM	
Dependency Ratio	: 30.5%

Part of the Total Renewable Water Resources originating outside the country expressed in percentage is called the dependency ratio of that country. India has a dependency ratio of 30.5 percent. Kuwait, Egypt, Turkmenistan, Bahrain, Mauritania, Sudan, Hungary, Bangladesh, Niger and Netherlands are top ten countries whose dependency ratios are more than 85 percent.

The water resources potential of the country (India) which occurs as natural run-off in the rivers is about 1869 BCM as per the estimates of Central Water Commission (CWC). A total storage capacity of about 213 BCM has been created in the country due to the major and medium projects since completed. The projects under construction will contribute to 76 BCM, while contribution expected from projects under construction is 107 BCM. Thus, likely storage available will be 396 BCM against a total availability of 1869 BCM in the river basins of the country. Major states like, Andhra Pradesh, Gujarat, Karnataka, Madhya Pradesh, Maharashtra, Orissa, and Uttar Pradesh together account for 70 percent of total live storage capacity in the country. The states of Arunachal Pradesh, Orissa, and Uttar Pradesh account for 72 percent of the total storage of projects under consideration. Uttar Pradesh, which lies mostly on Ganga river basin, has the highest ground water potential. The ground water potential of Uttar Pradesh is 81 BCM/yr., whereas, with 0.22 BCM/yr., state of Goa ranks as the lowest in terms of ground water potential. In terms of development of ground water, Haryana has the highest percentage followed by Punjab. Other states where percentage of ground water development is more than 50 percent are Gujarat (55 percent), Tamil Nadu (64 percent), and Rajasthan (86 percent).

Water resources potential in major river basins in India:

Sr. No.	Name of the River Basin	Average Annual Potential in the River	Estimated Utilisable Flow excluding Ground Water)
1	Indus (up to Border)	73.31	46.00
2	a) Ganga	525.02	250.00
	b) Brahmaputra, Barak and others	585.60	24.00
3	Godavari	110.54	76.30

Sr. No.	Name of the River Basin	Average Annual Potential in the River	Estimated Utilisable Flow excluding Ground Water)
4	Krishna	78.12	58.00
5	Cauvery	21.36	19.00
6	Pennar	6.32	6.86
7	East flowing Rivers between Mahanadi & Pennar	22.52	13.11
8	East Flowing Rivers between Pennar and Kanyakumari	16.46	16.73
9	Mahanadi	66.88	49.99
10	Brahmani&Baitarni	28.48	18.30
11	Subarnarekha	13.37	6.81
12	Sabarmati	3.81	1.93
13	Mahi	11.02	3.10
14	West Flowing Rivers of Kutch, Saurashtra including Luni	15.10	14.98
15	Narmada	45.64	34.50
16	Tapi	14.88	14.50
17	West Flowing Rivers from Tapi to Tadri	87.41	11.94
18	West flowing rivers from Tadri to Kanyakumari	113.53	24.27
19	Area of Island drainage in Rajasthan Desert	Neg	-
20	Minor River Basins drainage to Bangladesh & Myanmar	31.00	
	Total	1869.35	690.32

GROUND WATER STATUS IN INDIA

The state wise situation of ground water in India is enumerated as follows:-

Andhra Pradesh: The Annual Extractable Ground Water Resources in the state was 20.15BCM as estimated in 2017 and the Annual Ground Water Recharge during the same year was estimated at 21.22BCM which has increased from 20.39BCM since 2013. All watersheds have been assigned to respective mandals and ground water resources assessed accordingly. Out of the 670 mandals, 45 mandals have been categorized as 'Over Exploited', 24 as 'Critical', 60 as 'Semi Critical', and 501 mandals have been categorized as 'Safe'. 40 out of 670 mandals assessed, have been categorized as 'Saline'.

Arunachal Pradesh: The Annual Extractable Ground Water Resources for the state in 2017 was estimated at 2.667BCM and the Annual Ground Water Recharge during the same period was estimated at 3.025BCM. The water resources of the state was assessed district wise, leaving five districts namely, Upper Siang, Anjaw, Dibang Valley, Kurung Kumey and Tawang being in hilly areas of the state. There is no saline area in the state and all districts have been categorized as 'Safe'.

Assam: The total Annual Ground Water Recharge for Assam has decreased from 32.11BCM (Billion Cubic Metres) in 2013 to 28.67BCM in 2017. As per the report of the Department of Water Resources, the Annual Extractable Ground Water Resources was 24.26BCM in Assam in 2017. All 28 districts in the state have been categorized as 'Safe'.

Bihar: The total Annual Ground Water Recharge for the state has increased from 31.31BCM in 2013 to 31.41 BCM in 2017. As of 2017, the Annual Extractable Ground Water Resources was 28.99BCM. Out of the 534 blocks assessed in 2017, 18 were categorized as 'Critical', 12 as 'Over Exploited' and 432 blocks were considered 'Safe'.

Chhattisgarh: The total Annual Ground Water Recharge of the state was estimated at 11.57BCM and Annual Extractable Ground Water Resources was assessed as 10.57BCM in 2017. Out of the 146 blocks assessed, 22 blocks have been categorized as 'Semi critical', 2 as 'Critical' and 122 blocks have been categorized as 'Safe' blocks.

Delhi: The total Annual Ground Water Recharge as of 2017 has been assessed at 0.32 BCM and Annual Extractable Ground Water Resources was 0.30BCM. The total Ground Water Extraction for the state in 2017 was 0.36BCM. Out of 34 tehsils assessed, 22 tehsils have been categorized as 'Over Exploited' and 2 as 'Critical'. Only 3 out of 34 tehsils have been found 'Safe'.

Goa: The Annual Extractable Ground Water Resources of this state was 0.16BCM in 2017. The total Annual Ground Water Recharge in the same year was estimated at 0.27BCM. All of the 12 taluks assessed, have been categorized as 'Safe'.

Gujarat: The Annual Extractable Ground Water Resources for the state has increased from 19.79BCM in 2013 to 21.25BCM in 2017. During the same period, total Annual Ground Water Recharge has increased from 20.85BCM to 22.37BCM. As of 2017, 25 taluks out of 248 have been categorized as 'Over Exploited' and 5 as 'Critical'. 13 taluks have been categorized as 'Saline' in 2017 and 194 taluks in the state were considered 'Safe'.

Haryana: The Annual Extractable Ground Water Resources of Haryana was assessed at 9.13BCM, where as the total Annual Ground Water Recharge of the state has been assessed at 10.15BCM in 2017. There has been a ground water extraction of 137 percent. Out of the 128 blocks assessed, 78 blocks have been categorized as 'Over Exploited' and 3 as 'Critical' and 26 blocks were considered 'Safe'.

Himachal Pradesh: The Annual Extractable Ground Water Resources of the state in 2017 was estimated at 0.46BCM and total Annual Ground Water Recharge during the same year was 0.51BCM. Out of the 8 assessed units, 4 units have been categorized as 'Over Exploited'.

Jammu & Kashmir: The ground water resources were assessed in valley areas and outer plains of the 22 districts of the state in 2017. All assessment units of the state have been categorized as 'Safe'. The Annual Extractable Ground Water Resources was estimated at 2.60BCM in 2017 which has decreased from 4.82BCM in 2013. And the Annual Ground Water Recharge of the state has also decreased from 5.25BCM in 2013 to 2.89BCM in 2013.

Jharkhand: The Annual Extractable Ground Water Resources of the state in 2017 was estimated at 5.69BCM and total Annual Ground Water Recharge during the same year was 6. 21BCM. Out of the assessed 260 blocks, 3 blocks have been categorized as 'Over Exploited' and 2 as 'Critical'. 245 blocks out of 260 were considered 'Safe'. No saline blocks were found in the state. The Annual Extractable Ground Water Resources in Jharkhand decreased from 5.99BCM in 2013 to 5.69BCM in2017.

Karnataka: The Annual Extractable Ground Water Resources in Karnataka was 14.79BCM and the Annual Ground Water Resources as assessed in 2017 was 16.84BCM. Out of the assessed 176 taluks, 45 taluks have been categorized as 'Over Exploited' and 8 as 'Critical' in the state.

Kerala: The Annual Extractable Ground Water Resources of Kerala was estimated at 5.21BCM during 2017 and Annual Ground Water Recharge has been estimated at 5.77BCM. Out of the 152 blocks assessed, 1 has been categorized as 'Over Exploited' and 2 as 'Critical'.

Madhya Pradesh: The Annual Extractable Ground Water Resources in the state was assessed at 34.47BCM during 2017 and the Annual Ground Water Recharge has been assessed as 36.42BCM. As of 2017, out of the 313 blocks assessed, 7 blocks have been categorized as 'Critical' and 240 blocks have been marked 'Safe' with 44 as 'Semi Critical'.

Maharashtra: The Annual Extractable Ground Water Resources of the state in 2017 was estimated at 29.90BCM and total Annual Ground Water Recharge during the same year was 31.64BCM, which has decreased from the Annual Ground Water Recharge of 33.19BCM in 2013. Out of the 353 blocks assessed, 11 blocks have been marked as 'Over Exploited' and 9 as 'Critical'. 271 blocks out of 353 have been categorized as 'Safe'.

Manipur: Annual Extractable Ground Water Resources for the state has been estimated at 0.39BCM during 2017 and Annual Ground Water Recharge during the same year was 0.43BCM with a decrease of 0.04BCM since 2013. All districts of the state have been categorized as 'Safe'.

Meghalaya: The Annual Extractable Ground Water Resources in the state has decreased from 2.98BCM in 2013 to 1.64BCM in 2017 and the Annual Ground Water Recharge also decreased from 3.31BCM in 2013 to 1.83BCM in 2017.

Mizoram: The Annual Extractable Ground Water Resources for the state was estimated at 0.192BCM and the Annual Ground Water Recharge in 2017 was assessed as 0.213BCM. All the 26 blocks in the state has been categorized as 'Safe'.

Nagaland: The main source of domestic water supply in the state are the perennial springs in form of which ground water emerges. The Annual Extractable Ground Water Resources in 2017 was estimated at 1.98BCM and total Annual Ground Water Recharge was assessed as 2.20BCM. All the 11 districts in the state have been categorized as 'Safe'.

Orissa: Annual Extractable Ground Water Resources for the state has been estimated at 15.57BCM during 2017 and total Annual Ground Water Recharge during the same year was assessed as 16.74BCM. Out of the 314 blocks assessed in the state, 5 blocks were categorized as 'Semi Critical', 6 blocks as 'Saline' and rest 303 blocks have been categorized as 'Safe'. The state of ground water extraction has increased from 30 percent in 2013 to 42 percent in 2017.

Punjab: The state has three perennial rivers namely, Sutlej, Beas and Ravi. River Ghaggar is a non-perennial river. The Annual Extractable Ground Water Resources was estimated at 21.59BCM in 2017 and the total Annual Ground Water Recharge was assessed as 23.93BCM. There was annual ground water extraction of 35.87BCM in 2017 which was 66 percent above annual extractable ground water resources. The water resources in the state has been assessed block wise and out of 138 assessed blocks, 109 blocks have been categorised as 'Over Exploited', 2 as 'Critical', 5 as 'Semi Critical' and 22 blocks have been categorised as 'Safe'.

Rajasthan: The Annual Extractable Ground Water Resources of the state was estimated at 11.99BCM in 2017 and during the same period the total Annual Ground Water Recharge was assessed as 13.21BCM. As compared to this, the Annual Extractable Ground Water Resources of the state was 11.26 in 2013 and the total Annual Ground Water Recharge was assessed as 12.51BCM during the same year, with increase in both the estimated quantities. Out of the 295 blocks assessed, 185 blocks were categorised as 'Over Exploited', 33 as 'Critical', and 29 as 'Semi Critical'. Three blocks have been categorised as 'Saline' and 45 blocks were designated as 'Safe'.

Sikkim: The Annual Extractable Ground Water Resources of the state was estimated at 1.52BCM in 2017 and during the same period the total Annual Ground Water Recharge was assessed as 5.63BCM. There was insignificant annual ground water extraction in the state. All assessments units were categorised as 'Safe' with no ground water extraction reported in two districts of the state.

Tamil Nadu: The Annual Extractable Ground Water Resources was estimated at 18.20BCM in 2017 and the total Annual Ground Water Recharge was assessed as 20.22BCM. The ground water resources of the state have been assessed firka wise. Out of the 1166 firkas assessed, 462 have been categorised as 'Over Exploited', 79 as 'Critical', 163 as 'Semi Critical' and 35 firkas have been categorised as 'Saline'. Only 427 firkas out of 1166 assessed were categorised as 'Safe'.

Telengana: Like in Andhra Pradesh, all watersheds have been apportioned to respective mandals and ground water resources assessed accordingly in 2017. Out of the 584 mandals, 70 mandals have been categorised as 'Over Exploited', 67 as 'Critical', 169 as 'Semi Critical', and 278 mandals have been categorised as 'Safe'. No mandals were found to be 'Saline'. The Annual Extractable Ground Water Resources was estimated at 12.37BCM and the total Annual Ground Water Recharge was assessed as 13.62BCM during the same year.

Tripura: The Annual Extractable Ground Water Resources in the state as estimated in 2017 was 1.24BCM and Total Annual Ground Water Recharge was assessed as 1.53BCM. The ground water resources were assessed block wise and all the 39 blocks of the state have been categorised as 'Safe'. As compared to 2013, there was a decrease in both Annual Extractable Ground Water Resources and Total Annual Ground Water Recharge.

Uttar Pradesh: The Annual Extractable Ground Water Resources was estimated at 65.32BCM in 2017 and the Total Annual Ground Water Recharge of the state in the same year was assessed as 69.92BCM. Out of the 830 assessment units comprising of 820 blocks and 10 cities, 91 have been categorised as 'Over Exploited', 48 as 'Critical', 151 as 'Semi Critical' and 540 blocks were categorised as 'Safe'. No saline block was found in the state.

Uttarakhand: About 85 percent of the geographical area of this state is mountainous having underlying hard rocks. The ground water assessment in the state was done block wise. During the assessment year 2017, the Annual Extractable Ground Water Resources was estimated at 2.89BCM and the total Annual Ground Water Recharge was assessed as 3.04BCM. Out of the 18 blocks assessed, 5 blocks were found as 'Semi Critical' and 13 were found 'Safe'. No over exploited, critical or saline blocks were found in the state.

West Bengal: There were huge variations of results in ground water assessment of the state in 2017 with those of previous assessment in 2013 and reasons for changes provided were not tenable and adequate as per report. And thus, 2013 assessment results were considered, when the Annual replenishable resources had been estimated at 29.33BCM and Net Ground water availability as 26.56BCM.

Irrigation Potential: The total ultimate irrigation potential of the country is approximately 140mha. Through minor irrigation, ground water contributes more than 79 percent of total ultimate irrigation potential. The largest ultimate irrigation potential (UIP) for minor irrigation through ground water, exists in Uttar Pradesh. Uttar Pradesh occupies the first place among all states to have maximum potential considering all types of schemes. In terms of arable land, United States of America has the highest arable land comprising 173.5 mha, with India occupying second place and having an arable land area of 160.5 mha.

RAIN THE PRIMARY SOURCE

Rain is the primary source of all water on earth. Rain by definition is a form of precipitation which contains droplets of water having a size bigger than 0.5mm and which forms as a process of condensation of clouds in the atmosphere. Precipitation is any form of water that falls from the clouds in the sky.

Types of rain:
There are three basic forms of rain. They are:
1. *Convectional rain*: In hot and humid places or in the inland areas with hot climatic conditions, when the temperatures are high during the summer, the air near the surface of the earth gets heated due to conduction and expands due to the heat and its density reduces. In the process, due its lower density, this air gets lifted in to the sky. As the rising air goes up, its temperature goes down and falls to the dew point forming water vapour and clouds in the sky. Depending on the temperature, the clouds get converted in to rain and falls as sleet or snow. This is called the convectional rain.
2. *Relief rain*: When the moist wind blowing from the sea towards a coastal region suddenly comes across a relief feature on the land

mass i.e. a hill or a mountain. The wind rises up or is forced to rise up along the windward slope. As the air rises up along the wind facing slope of the elevated land, its temperature falls and condenses to form clouds. The water droplets are then formed due to further decrease in temperature in the clouds which cannot hold these water droplets any further and thus fall as rain. This is the relief rain. The opposite side of the hill or mountain remains devoid of any rain or with scanty rain fall. This side of the hill or mountain is called the leeward side.

3. *Frontal rain*: The border region between two adjacent air masses having different temperature and humidity is called a front. When, warm and moist air travelling from one direction meets cold air front travelling from opposite direction, the cold air under run the warm air in form of a flat wedge. The warm air being less dense and lighter rises along and above the cold air front boundary. As the warm air ascends, its temperature decreases and results in formation of clouds which when rises further gets heavier and falls as rain. This is called the frontal rain.

Measurement of Rainfall: The amount of rain is expressed as the depth in centimeters which falls on a level surface and is measured with a rain-gauge. The most common type of non-automatic rain-gauge used to measure amount of rain is the Symon's rain-gauge. It consists of a cylindrical vessel of 127 mm (5") diameter. The base of the vessel is enlarged to 210mm (8") diameter to give stability. The cylindrical vessel houses a receiving bottle of diameter 75mm (3") to 100mm (4"). A funnel whose size matches the diameter of the cylindrical vessel is inserted in to the receiving bottle so that rain drops falling inside the vessel reaches the receiving bottle through this funnel. The receiving bottle of the rain-gauge has a capacity to measure 75mm to 100mm of rainfall. If the rainfall is heavy or extreme, the measurement is taken 3 or 4 times a day so that the receiving bottle does not overflow. The rain-gauge is installed in a concrete block of size 60cm x 60cm x 60cmm. A cylindrical graduated measuring glass is furnished along with rain gauge which can read nearest to 0.2mm. The location of the rain-gauge should be such that:

- It should be set up in a level and an open place.
- The distance between the rain-gauge and the nearest object should at least be twice the height of the object.
- It should be at least 30 metres away from the nearest obstruction.
- As far as possible it should not be set up on top of a hill so that high speed winds are avoided.
- It is preferable to erect one fence around the rain-gauge station so that it is protected from cattle etc.

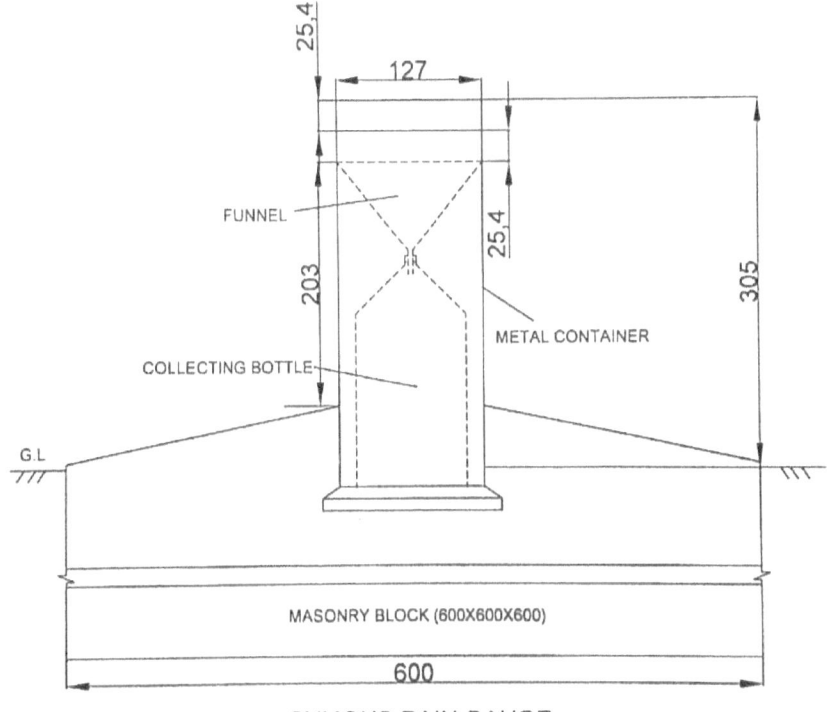

SYMONS RAIN GAUGE

(ACAD representation: Author)

The commonly used automatic rain-gauge is the Float type automatic rain gauge which contains a funnel to receive the rain water. The rain water through the funnel reaches this receiving container which houses one float connected to a float rod with pen arm which helps to measure the quantity of rainfall directly from the recording drum. When the level of water in the receiving container rises, the float also rises and correspondingly the pen arm of the float rod records the amount of rainfall from the recording drum.

Computation of Average rainfall: To get an even picture of the average rainfall over a catchment area, the network of rain-gauge stations are so distributed across the catchment that they give a representative picture of the entire area under consideration. In general, following guidelines can be followed based on the area of catchment.

Sl.No.	Area in Square Km			Number of rain-gauge stations
1	0	to	80	1
2	80	to	160	2
3	160	to	320	3
4	320	to	560	4
5	560	to	800	5
6	800	to	1200	6

Average rainfall is found out using following methods:

(i) *Arithmetic Average method*: The recorded rainfall values of all the rain-gauge stations are noted and the average of the rainfall values of the area is found out by adding all values and dividing the sum so worked out by the number of rain-gauge stations. Thus

$$P_{av} = (P_1 + P_2 + \ldots\ldots\ldots\ldots Pn)/n,$$

Where $P_1, P_2, P_3 \ldots Pn$ are the rainfall values of individual rain-gauge stations falling under the catchment area and 'n' is the number of such rain-gauge stations.

(ii) *Thiessen Polygon method*: Thiessen Polygon method of finding the average rainfall is more reliable than the Arithmetic Average method since the rainfall varies from place to place in terms of quantity and duration and taking average may not give a true picture. The rainfall from a particular location is weighed according to the area it represents. Thus, following procedure is adopted in this method.

A map of the entire catchment is drawn. The places where rain-gauges are stationed (say 1, 2.... n) are joined by straight lines. Then perpendicular bisectors to these lines joining rain-gauge stations are drawn. Area demarcated by these perpendicular bisectors and catchment boundaries will represent the individual rain-gauge stations. The average rainfall is given by-

$P_{av} = (A_1P_1 + A_2P_2 + \ldots\ldots\ldots\ldots A_nP_n)/ (A_1+A_2+\ldots\ldots A_n)$,

Where-

$A_1, A_2,\ldots\ldots A_n$ represent respective areas enclosed within the perpendicular bisectors and catchment boundaries.

$P_1, P_2,\ldots\ldots P_n$, represent the precipitation or rainfall values at corresponding rain gauge stations 1, 2……n.

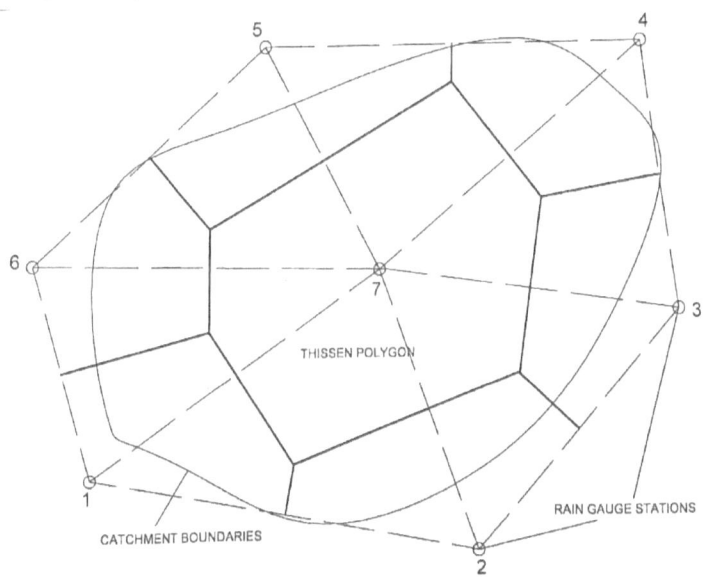

(ACAD representation: Author)

Intensity of rainfall: The intensity or amount of rainfall is measured with the help of a rain gauge. The rain is classified according rate of precipitation as:

(i) *Very light rain*: When the rainfall has a precipitation rate of less than 0.25 millimeters per hour. It is classified as Very light rain.

(ii) *Light rain*: When the rainfall has a precipitation rate between 0.25 millimeters per hour and 1 millimeter per hour. It is classified as Light rain.

(iii) *Moderate rain*: Rainfall with a precipitation rate between 1 millimeter per hour and 4 millimeters per hour is classified as Moderate rain.

(iv) Heavy rain: When the intensity of rain or the rate of precipitation is between 4 millimeters per hour and 16 millimeters per hour, it is classified as Heavy rain.

(v) Extreme rain: Rainfall with precipitation rate more than 50 millimeters per hour is termed as Extreme rain.

Monsoon: The word "monsoon" has its origin in the Arabic word "mausim" which means season. Since several centuries ago, the fisher men and sailors at sea noticed winds which seem to come from north east for six months and again from the south west for another six months. They used to refer these seasonal winds as mausim. There is no precise definition of monsoon other than referring this as alternating seasonal winds which blows consistently and with regularity for a period of the year from one direction and from another direction for another part of the year. Climatic conditions which favour onset of these winds is better known as the monsoon climate. The continents of Asia, Africa, and Australia are covered by monsoon climate. The boundaries of monsoons lie between 35°N & 25°S and 30°W & 173°E. During the summer in northern hemisphere, the south westerly monsoon blows over India, South China, South East Asia and West Africa. The monsoon in India during summer in northern hemisphere is known as the "south west monsoon" or the "summer monsoon". The monsoon during winter in northern hemisphere over the south eastern part of Indian peninsula is referred to as "northeast monsoon". Winter rain in Tamil Nadu and Andhra Pradesh owes its presence to the onset of this northeast monsoon.

Theories and Hypothesis associated with occurrence of monsoon: In 1686, Edmund Halley hypothesized the occurrence of monsoon to a planetary scale "land breeze-sea breeze" system. Halley propounded that the occurrence of monsoon is due to differential heating of land and ocean associated with the seasonal movement of the sun. In summer, when the land mass is heated up, a pressure difference in atmosphere between the land and sea is created thereby the cool breeze from the adjacent sea moves in to the land mass. And in winter, when the continents cool to a temperature lower than the sea, there is a reverse flow of wind from the continents to adjacent sea or ocean. Circulation of wind in case of monsoon is similar to the 'land breeze' and 'sea breeze' except that the day and night in case of land breeze and sea breeze is replaced by summer and winter season and the narrow strip of land and the adjacent sea is replaced by large portion of continents and the oceans.

Halley's model did not take in to account the swirling motion of the wind while blowing from ocean to the land mass of the continents. His theory only suggested that the wind would blow directly from sea to the coasts. However, in 1735, George Hadley noted that due to the rotation of the earth around its axis. The winds will be affected by the coriolis force. The coriolis force deflects any motion to the right in the northern hemisphere and motion to the left in the southern hemisphere. Thus, moving winds of the Halley's model undergoes deflection to the right due to the coriolis force in the northern hemisphere and deflection to the left in the southern hemisphere, thus causing north easterlies and south westerlies.

Onset and withdrawal of monsoon in India: In the later part of the month of May, the south east trade winds moving towards north on crossing the equator gets deflected due to coriolis force and becomes the south westerlies which continue to move towards north and reach Sri Lanka by last week of May and thus Sri Lanka gets its first shower of monsoon rain by last week of May. The south westerlies on reaching Sri Lanka divide in to two branches- Arabian Sea branch and the Bay of Bengal branch. Further moving in northern direction, the monsoon reaches the southernmost tip of Indian peninsula by 1st week of June. The Arabian branch of the monsoon moves north ward and reaches Mumbai by 10th June. The Bay of Bengal branch moving towards north spreads over parts of Assam by 1st week of June. Further the Bay of Bengal branch moves towards north and reaches the southern boundary of Himalayan mountain ranges and gets deflected towards the west and move towards Gangetic plains and reaches Kolkata by 7th June. The Arabian branch of the monsoon spreads over most of Saurastra-Kutch region and central India. By middle of June the deflected branch of the Bay of Bengal and the Arabian branch of monsoon merge in to a single current. Remaining part of western Rajasthan experiences monsoon by 1st of July. Kashmir experiences monsoon only as a feeble current by mid-July. Beginning 1st of June, the monsoon in India lasts for about 100 to 120 days. By end of October, the monsoon withdraws from the North

West India and withdraws from the rest of the country by December. India Meteorological Department has 14 weather stations located in Lakshadweep, Kerala & in Mangalore. The rainfall recorded in these weather stations are monitored from May 10, if the 60% of the stations or more record more than 2.5mm of rainfall for two continuous days, then it satisfies the criteria that monsoon has arrived.

Monsoon and India: The rain associated with monsoon also called the Indian Summer Monsoon Rain (ISMR) has high spatial variability. The Western Coast of peninsular India and country's North Eastern Region receiving higher amount of rainfall exhibits lesser spatial variability compared to North Western parts of India receiving lower amount of rain and having a larger spatial variability. Indian Summer Monsoon Rain (ISMR) has special significance in Indian economy as the same has direct or indirect effect on agriculture, water resources, transportation, health, power sectors. Livelihood of vast majority of the rural population is affected by a deficient ISMR. Most of the regions in peninsular central and north western regions of the country are prone to drought. These areas receive less than 100cm of rainfall annually. Again, a vast majority of approximately 68 percent of the cropped land in the country falls under regions receiving low to moderate ISMR. Therefore, one or two successive years of difficult monsoon can create havoc. The drought in 2002, 2004 and 2009 has affected all the regions of low rainfall. The drought in 2002 forced 10 states in the country to place a demand of Rs. 11,000 Crores from Centre even by August '2002, details of which are given as under:-

State	Rainfall	Damage	Demand for aid from centre (Crores)
Punjab	-49%	7 of 17 districts affected badly	800
Uttar Pradesh	-59%	58 out of 70 districts affected	750
Rajasthan	-67%	41,000 villages in 32 districts hit	4996
Chhattisgarh	-38%	All 16 districts drought hit	1700
Haryana	-69%	60%-90% crops damaged	650
Karnataka	-40%	Rs. 1700 Crores worth of crops affected	533
Andhra Pradesh	-39%	58 lakhs farm workers hit	610

Source: India Today, August 12, 2002.

The drought in year 2002 caused a reduction of Food Grain Production (FGP) of 13 percent; while the drought in 2009 has erased the country's GDP by about 0.5 percent. India receives about 80% of monsoon rainfall during the four months from June through September. The average rainfall for the country is 88cm. The highest rainfall recorded till this date is 1187cms at Mowsinram near Cherrapunji in the state of Meghalaya. The lowest recorded rainfall is 0.011cm at Jaisalmer in Rajasthan. The monsoon is considered normal if it is +/- 10% of its long-term average. During the period from 1901 to 2011, the lowest seasonal rainfall occurred in 1918 and 1972 with deficient rainfall of 75.1 and 76.40 percent of the long-term average respectively, the highest seasonal rainfall occurred in 1917 and year 1961 with excessive rainfall of 122.90 percent and 121.80 percent of the long-term average respectively. Out of 20 drought years that occurred between 1901 and 2011, 13 years (65 percent) coincided with El Nino events. But there has been long span of normal monsoon seasons too in the past, lasting even up to twenty years, like from 1921 to 1940, where every monsoon season has been normal. The country on an average receives an annual precipitation of about 4000BCM (Billion Cubic Metres); this includes the precipitation in form of snowfall. For the purpose of recording seasonal rainfall, the country has been divided in to 36 meteorological sub divisions. The country recorded a rainfall volume of 4057BCM during the year 2003, 3570BCM in 2004 and 3972BCM during 2005.

Monsoon Prediction: Monsoon rains are predicted for three ranges of time periods (i) Short range – (1-3 days) (ii) Medium range- (4-10 days) (iii) Long range Forecast – More than 10 days to a season or beyond.

Short and Medium range of forecasting: Short and Medium range of forecasting of monsoon rains is done using Numerical Weather Predicting (NWP) models. The state of the atmosphere at a given time is noted and using fluid-dynamics and thermodynamics, state of the atmosphere some time in future is estimated. The inputs to these models are the surface observations from weather stations over the land; weather buoys at sea, observations at different heights of atmosphere obtained by specialized instruments flown in to atmosphere with hydrogen balloons. Where traditional data sources are not available, data from the weather satellites are taken for those areas. These observed data are used as initial inputs to these models. The data are further assimilated to find their values at evenly spaced grids so that these are usable by model's mathematical algorithms to find the estimated values at certain interval in future.

Long Range Forecasting: The first Chief Reporter of the India Meteorological Department (IMD) Sir H.F.Blanford issued tentative forecasts for monsoon rains from 1882 to 1885 based on indications provided by snowfall in Himalayas. The regular forecasts for seasonal monsoon rainfall during June to September started from 4th June 1986. The efforts to forecast monsoon rain continued during the tenure of Sir Gilbert.T.Walker from 1904 to 1924. In 1988 IMD introduced the 16-power regression and parametric model and started issuing forecasts for the country as whole. The model had an error of +/-4% and was giving satisfactory results till 2001. However, with failure of the model to predict ISMR in 2002, IMD introduced a new two stage strategy in 2003. The seasonal forecast for the country as a whole is issued once in April and another update forecast in June. A set of 8(eight) predictors, having stable and strong physical linkages with ISMR, is used. Details of the eight predictors derived from global ocean-atmosphere parameters and used for statistical forecasting-

Sl.No	Predictor	Used for forecasts in	Correlation Coefficient
1	NW Europe Land Surface Air Temperature(P1)	April	0.58
2	Equatorial Pacific Warm Water Volume(P2)	April	-0.30
3	North Atlantic Sea Surface Temperature(P3)	April and June	-0.49
4	Equatorial SE Indian Ocean Sea surface Temperature(P4)	April and June	0.45
5	East Asia Mean Sea Level Pressure(P5)	April and June	0.36
6	Central Pacific (Nino3.4)Sea Surface Temperature Tendency(P6)	June	-0.49
7	North Atlantic Mean Sea Level Pressure(P7)	June	-0.52
8	North Central Pacific wind at 1.5 Km above sea level(P8)	June	-0.44

For April forecast, the first 5 predictors are used and for June, last 6 predictors are used which include 3 predictors used for April forecast.

Ministry of Earth Sciences (MoES) has launched the National Monsoon Mission for developing a sophisticated and reliable dynamical model for prediction of ISMR at different time frame with Indian Institute of Tropical Meteorology (IITM), Pune coordinating with various climate research agencies in India and abroad. The Monsoon Mission would use two coupled ocean- atmosphere models, (i) the Coupled Forecast System (CFS) version 2.0 of National Centre for Environmental Prediction (NCEP) and (ii) the Unified Model (UM) of U.K. Met office.

WATER POLLUTION AND EFFECT ON HEALTH

Polluted water contains various metallic ions, e.g., arsenic, molybdenum, lead, cadmium, mercury, nickel, cobalt etc. many of these are highly toxic and when present in river or pond water effect plant and animal life.

In India, all the 14 major rivers including Coorum, Ganga, Gomti, Cauvery, Damodar and Mini Mahi have been polluted. Most of the rivers in India flowing near cities and towns are polluted. The river Damodar has very low dissolved oxygen and cannot therefore support any growth of useful aquatic fauna and flora. It is highly infested with pests and pesticides. Same is the case with river Mini-Mahi in Baroda; the river contains a variety of industrial and petrochemical wastes. It is said that river Cooum, flowing through Chennai city, is so much polluted by sewage that not even the Zooplanktons are able to survive. The sulfate levels in the river is said to be highest among Indian rivers. The Oudh Sugar mills of Birla's and a distillery of Mohan Meakins are believed to be responsible for polluting a stretch of 8Kms of river Gomti, which is responsible for death of thousands of fishes. The mercury released from a factory near Mettur Dam has polluted river Cauvery in Tamil Nadu. Urban liquid wastes, Carcasses of animals, tons of pesticides and insecticides fill river Ganga from city of Hardwar to Kolkata. In its journey through nearly 1500 miles to Bay of Bengal, 27 towns and cities dump 905 million litres of wastewater every day affecting the health of some 251million people in North India.

The wastewater from distilleries contains high BOD, with brownish colour and pungent odour. It also contains high content of dissolved solids. The wastewater from acid manufacturing industries contains low pH. The wastewater from paint industries contains high BOD, also carries synthetic resins, solvent pigments, and heavy metals such as Chromium and Lead. Petroleum industries release wastewater with high COD/BOD ratio and also contains hydrocarbons, alcohols, aldehydes, phenols, oils, metals etc. Plastic manufacturing units release wastewater having trace of acids, formaldehydes and phenols. Wastewater released from pulp and paper industries contains high dissolved and suspended solids. Steel industries release wastewater containing low pH and impurities such as phenols and metals. Wastewater from tanneries contains highly dissolved solids, oil and grease, heavy metals like Chromium. Higher amount of suspended solids, NH_3 and H_2S is found in wastewater released from coke manufacturing industries.

Some polluted rivers in India and their sources of pollution:

Name of River	Source of pollution
Bhadra (Karnataka)	Pulp, Paper and Steel Industries.
Cauvery (Tamil Nadu)	Sewage, Tanneries, Distilleries, Paper and Rayon mills.
Chambal (M.P)	Rayon mills, Caustic Soda.
Cooum (Tamil Nadu)	Automobile workshops.
Damodar (Bakaro)	Fertilizers, Fly-ash from Steel mills
Ganga (At Kanpur, U.P)	Jute, Chemical, Metal and Surgical industries; Tanneries, Textile mills, domestic sewage of highly organic nature.
Godavari (A.P)	Paper mills.
Gomti (Near Lucknow, U.P)	Sewage, Paper and Pulp mills.
Hooghly (Near Kolkata)	Power stations; Paper, Pulp, Jute, Textile mills; Chemical, Paint, Varnishes, Metal, Soap, Polythene industries.
Jamuna (Near Delhi)	DDT factory, Sewage.
Kali (At Meerut, U.P)	Sugar mills, Distilleries, Paint, Soap, Rayon, Silk, Yarn, and Glycerine industries.
Kulu (H.P)	Chemical factories, Rayon mills and Tanneries.
Narmada (M.P)	Paper mills.
Siwan (Bihar)	Sulphur and Sugar mills; Cement industries.

Pesticides and Biocides

Pesticide means any substance intended for preventing, destroying, repelling or control of any pest including unwanted species of plants and animals during production, storage, transport and distribution and processing of food, agricultural commodities or animal feeds. Pesticides are used to ensure better protection at harvest against unpredictable losses caused by plant diseases and pests. They are used to improve both quality and quantity of food. Pesticides are used to decrease the extent of vector borne and other diseases in humans and animals. However, these chemicals create problems when they are present in high concentration. Some of the banned pesticides in India are: 1. Dibromochloropropane(DBCP), 2. Endrin, 3. Pentachlornirobenzene(PCNB), 4. Pentachlorphenol (PCP), 5. Toxaphene, 6. Ethyl Parathion, 7. Chlorodane, 8. Heptachlor, 9. Aldrin, 10. Nitrofen, 11. BHC, 12. Tetradifen.

Source and Toxic effects of some pollutants:

Sl.No.	Name of Compound/ Pollutant	Source	Effect on Health
1	Vinyl Chloride	Used in Plastic	Damage to liver, bone and circulatory system, Cancer of liver, brain, and lymphatic system.
2.	Benzene	Used in Detergents, Mouldings, Insecticides.	Anaemia, Leukemia.
3.	Aldrin / Dialdrin	Insecticides	Cause tremors, Convulsions, Damage to kidney.
4	DDT	Insecticides	Causes tremors, Degradation of central nervous system.
5	Dioxin	Herbicide	Powerful carcinogen, Causes chromosome malformation.
6	Nitrates & Nitrites	Septic tanks, Heavily fertilized crops, and Sewage treatment plants	Nitrates combine with haemoglobin to form Methaemoglobin which interferes with the oxygen carrying capacity of the blood, producing a serious disease known as Methaemoglobinaemia.

DDT: DDT was first made in 1874 by a German chemist Ziedler, then, as an insecticide in 1939 by a Swiss chemist Paul Muller. It is a white amorphous powder with a mild but not unpleasant smell. DDT shows a phenomenon called biological magnification. It means there is continuous increase in concentration of these harmful substances in successive trophic level or resident hosts. For instance, the DDT spread in water may be low, but the algae and some small organisms living in such conditions show an average of about 5.5ppm, a 265-fold increase over initial application of DDT. The herbivores, fishes and insects thriving on these plants may accumulate this chemical in still higher concentration, and by the time, the food chain reaches the top carnivores, the concentration of DDT in such carnivores like some shell fish (e.g. Oyster) may reach more than 70,000 times than its original application concentration in water. A classic example of biological magnification of DDT in food chain is; spraying on Elm trees resulted in 99ppm of DDT concentration in the soil, 140ppm in earthworm and more than 400ppm of DDT concentration in robins that feed on these earthworms. The biological magnification of DDT from water to fish eating gulls (birds) has been found to increase from mere 7ppb to 100ppm, an increase of 14,286 times.

Mercury: Mercury is one of the most dangerous pollutants among naturally occurring or an industrial pollutant. The first major disaster of mercury poisoning occurred in Minamata city of Japan in 1953, with at least 43 reported deaths due to eating of contaminated fish. Since then it has been termed as Minamata disease. The mercury was released in form of methyl mercury that came with industrial waste water from Chisso Corporation's chemical factory producing Vinyl Chloride, where mercury chloride was used as a catalyst. Also, mercury used as a seed dressing chemical to prevent fungus infection has resulted in numerous instances of bird and wild life poisoning. The symptoms of Minamata disease involved malaise, numbness, visual disturbance, dysphasia, ataxia, mental deterioration, convulsion and finally death of affected persons. In Kerala (India), indiscriminate use of mercury in gold extraction and rayon manufacture has polluted Chaliyar River. A large number of people have been affected due their exposure to contaminated water from the river. The mercury content of the river water was found to vary between 0.007ppm to 2.9ppm as against the prescribed limit of 0.001ppm.

Fertilizers: Excessive and indiscriminate use of chemical fertilizers and synthetic feed for livestock often leads to accumulation of nitrates in water. When cattle and humans drink such waters, these nitrates taken in to body are converted to toxic nitrites by intestinal bacteria. These nitrites combine with haemoglobin to form methaemoglobin, which reduces oxygen carrying capacity of the blood and produces a disease called methaemoglobinaemia. The various ailments that result from this disease include, damage to respiratory and vascular system, blue colouration of the skin and even cancer. A healthy person has 0.8 percent of methaemoglobin. With symptoms of methaemoglobaminaemia becoming conspicuous with changing of skin colour and mucus, the methaemoglobin content in such affected person can reach up to 10 percent.

Thermal pollution: Chemical industries, fossil fuel and nuclear power plants use lot of water for cooling purposes and return this water to stream at a higher temperature. The hot water interferes with the natural conditions in the lake and river affecting aquatic life. This is thermal pollution as heat acts as a pollutant. The thermal pollution is thus the raising of temperature of part of the environment (water in this case) by the discharge of substances whose temperature is higher than the ambient temperature. The water, if very hot, kills some plants and animals instantly. Some of the adverse environmental effects of the thermal pollution are: 1. Fish eggs hatch much early in the spring due to which the natural food organisms for young offspring would not be available. 2. Salmon often fail to spawn. 3. BOD increases as warm water holds less Oxygen. Water at 0.0°C has approximately 14ppm of dissolved oxygen, whereas at 20.0°C has only 9ppm. 4. Change in diurnal and seasonal behaviour and metabolic responses of organisms. 5. Significant shift in type algae and other organisms towards more heat-resistant species. 6. Affects migration of some aquatic life.

Government of India has set some guidelines regarding quality of drinking water through public water supply to consumers. As per Bureau of Indian Standards specifications, prescribed Drinking Water Quality Standards in India as per IS 10500-2012 are as under:

Sl. No.	Characteristic	Unit	Requirement (Acceptable limit)	Permissible limit in the absence of alternate source
1	Total Dissolved Solids (TDS)	Mg/L	500	2000
2	Colour	Hazen Unit	5	15
3	Turbidity	NTU	1	5
4	Total Hardness	Mg/L	200	600
5	Ammonia	Mg/L	0.5	0.5
6	Free Residual Chlorine	Mg/L	0.2	1
7	pH	—	6.5-8.5	6.5-8.5
8	Chloride	Mg/L	250	1000
9	Fluoride	Mg/L	1.0	1.5
10	Arsenic	Mg/L	0.01	0.05
11	Iron	Mg/L	0.3	0.3
12	Nitrate	Mg/L	45	45
13	Sulphate	Mg/L	200	400
14	Selenium	Mg/L	0.01	0.01
15	Zinc	Mg/L	5.0	15.0
16	Mercury	Mg/L	0.001	0.001
17	Lead	Mg/L	0.01	0.01
18	Cyanide	Mg/L	0.05	0.05
19	Copper	Mg/L	0.05	1.5
20	Chromium	Mg/L	0.05	0.05
21	Nickel	Mg/L	0.02	0.02
22	Cadmium	Mg/L	0.003	0.003
23	E-Coli or Thermotolerant coliforms	Number in 100 ml	Nil	Nil

*Mg/L: Milligram per litre.

Further, World Health Organisation (WHO), has provided the following guidelines relating to few contaminants / chemicals present in drinking water supply; These WHO's guidelines for Drinking-water Quality, set up in Geneva, 1993, are the international reference point for standard setting and drinking-water safety.

Sl. No.	Element/Substance	Health based guidelines by WHO
1	Aluminium	0.2 mg/l
2	Antimony	0.005 mg/l
3	Arsenic	0.01 mg/l
4	Barium	0.3 mg/l
5	Cadmium	0.003 mg/l
6	Chloride	250 mg/l
7	Chromium	0.05 mg/l
8	Copper	2 mg/l
9	Cyanide	0.07 mg/l
10	Fluoride	1.5 mg/l
11	Lead	0.01 mg/l
12	Manganese	0.5 mg/l
13	Mercury	0.001 mg/l
14	Nickel	0.02 mg/l
15	Nitrate	50 mg/l
16	Selenium	0.01 mg/l
17	Sodium	200 mg/l
18	Sulphate	500 mg/l
19	Zinc	3 mg/l

Chapter 2

CRISIS

10 (ten) facts about water scarcity as listed by W.H.O:
1. Water scarcity occurs even in areas where there is plenty of rainfall or freshwater. How water is conserved, used and distributed in communities and the quality of the water available can determine if there is enough to meet the demands of households, farms, industry and the environment.
2. Water scarcity affects one in three people on every continent of the globe. The situation is getting worse as needs for water rise along with population growth, urbanization and increases in household and industrial uses.
3. Almost one fifth of the world's population (about 1.2 billion people) lives in areas where the water is physically scarce. One quarter of the global population also live in developing countries that face water shortages due to a lack of infrastructure to fetch water from rivers and aquifers.
4. Water scarcity forces people to rely on unsafe sources of drinking water. It also means they cannot bathe or clean their clothes or homes properly.
5. Poor water quality can increase the risk of such diarrhoeal diseases as cholera, typhoid fever and dysentery, and other water-borne infections. Water scarcity can lead to diseases such as trachoma (an eye infection that can lead to blindness), plague and typhus.
6. Water scarcity encourages people to store water in their homes. This can increase the risk of household water contamination and provide breeding grounds for mosquitoes - which are carriers of dengue fever, malaria and other diseases.

7. Water scarcity underscores the need for better water management. Good water management also reduces breeding sites for such insects as mosquitoes that can transmit diseases and prevents the spread of water-borne infections such as schistosomiasis, a severe illness.
8. A lack of water has driven up the use of wastewater for agricultural production in poor urban and rural communities. More than 10% of people worldwide consume foods irrigated by wastewater that can contain chemicals or disease-causing organisms.
9. Millennium Development Goal number 7, target 10 aimed to halve, by 2015, the proportion of people without sustainable access to safe drinking water and basic sanitation.
10. Water is an essential resource to sustain life. As governments and community organizations make it a priority to deliver adequate supplies of quality water to people, individuals can help by learning how to conserve and protect the resource in their daily lives.

In general, a country with per capita availability of less than 1700 cubic metre is considered as 'water stressed' nation and a country with less than 1000 cubic metre of per capita availability are regarded as 'water scarce'. The projection of the future needs and the availability of fresh water resources indicates that by year 2025, 48(forty-eight) countries with more than 2.8 billion people covering nearly 35% of the projected world population will be affected by water scarcity. And by year 2050, fifty-four countries a population of approximately 4 (four) billion will face water scarce condition.

In 1955, there were only three Middle East countries falling under water scarce category. They are- Bahrain, Jordan and Kuwait. However, by 1990, another eight countries joined the list. They are- Algeria, Israel, Qatar, Saudi Arabia, Somalia, Tunisia, UAE and Yemen. A UN study has predicted that another ten countries would join the list by 2025. Statuses of some of the Middle East countries with their annual renewable fresh water per capita availability with projected per capita availability in 2025 are given below:

Sl.No.	Country	1955	1990	2025
1	Kuwait	147	23	9
2	Qatar	808	75	57
3	Bahrain	1427	117	68
4	Saudi Arabia	1266	306	113
5	UAE	6196	308	176
6	Jordan	906	327	121
7	Israel	1229	461	264
8	Libya	4105	1017	359
9	Egypt	2561	1123	630
10	Iran	6203	2203	816
11	Iraq	18441	6029	2356

(Adel Darwish, Geneva conference on "Environment and Quality of life" – June'1994.)

India occupies roughly 2.4 percent of the world's land mass and has 4 percent of world's water resources. But the Indian population being approximately 16% of the total population on earth, reduces the per capita availability. Again, of the 4 percent available water resources, nearly 66 percent of these are found in three major river basins of Ganges, Brahmaputra and Godavari which covers an area of only 33 percent. Brahmaputra and Barak basin having a geographical area of 7.7 percent and population of 4.2 percent of all the river basins of the country has 31 percent of annual fresh water resources. Cauvery, Pennar, Sabarmati are some of the basins where per capita availability is less than 1000 cum and thus fall under water scarce category. It is estimated that, India's per capita availability of fresh water resources in year 1975 was about 25.7 percent of average world availability. This has dropped to 23.4 percent of world availability in year 2000 and expected to decline further to about 23 percent in the year 2025. Past population and future projections and corresponding per capita availability of the country are given as under:

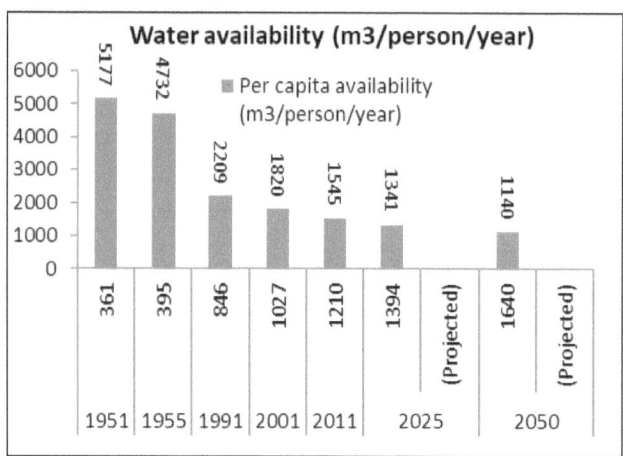

From figures above, it is now clearly evident that India is steadily sliding down to become a water scarce nation and with its population of 1.21 billion in the year 2011 and a corresponding per capita availability of 1545m³, and with a population of 1.37 billion in 2020, it is already a nation that is considered water stressed by UN classifications.

Water withdrawal: Ogallala aquifer is considered the largest single water bearing unit in entire North America. Stretching some half a million square kilometers in American High Plains from Texas panhandle to South Dakota, the Ogallala aquifers is believed to contain up to 4 trillion cubic meters of fossil water. The deep underground water that has been there since thousands of years with little replenishments is presently being mined mercilessly with some 200,000 wells to irrigate approximately 3.3 million hectares of American farm land. At withdrawal rate of 50 million liters per minute, the Ogallala aquifer is being depleted at a rate 14 times faster than the nature can restore it. Since 1991, the water level in Ogallala aquifer has gone down by at least one-meter due excessive withdrawal. It is believed that the once great reserve of fossil water, the Ogallala aquifer has already lost more than half of its water reserves.

In developing countries, ground water is widely used and forms a major portion of supplies to the citizens. The ground water is primarily used because of their low cost and usual high quality. However, over-abstraction of ground water has depleted ground water aquifers in many cities around the world. In Brazil, Manila and Chinese city of Tianjin, the water level has fallen by as much as 20m to 50m and in many others by 10m to 20m. In all these cases, drop in levels have been accompanied by subsidence or deterioration in ground water quality or both. The Mexico City saw a fall of its aquifers' levels by 5m to 10m between 1986 and 1992 and a section of the city have sunk by 8m or more during the last 60 years. In coastal belts, the over exploitation of ground water sources resulted in salt water intrusion. In Europe, 53 out of 126 ground water areas show salt water intrusion. In India, the ground water table in 203 districts of the 593 already reached a critical level even by the year 2003. Fresh water withdrawals in agriculture exceed the withdrawal for other sectors such as house hold & manufacturing etc. 44 percent of all water withdrawal in Organization for Economic Co-operation and Development (OECD) countries is attributed to agriculture.74 percent of all water withdrawal in BRIC countries (Britain, Russian Federation, India, China), is attributed to agricultural sector. In Least Developed Countries (L.D.C), the water withdrawal for agriculture accounts for more than 90 percent of all water withdrawal. In India, the water for agriculture accounts for 87 percent of the total withdrawal. The water withdrawal in developing countries is expected to rise by 50 percent and in developed countries by 18 percent by the year 2025. Just ten countries, India, China, U.S.A, Pakistan, Iran, Bangladesh, Mexico, Saudi Arabia, Indonesia and Italy accounted for about 72 percent of all ground water abstraction on earth. The ground water abstractions by these countries in 2010 are

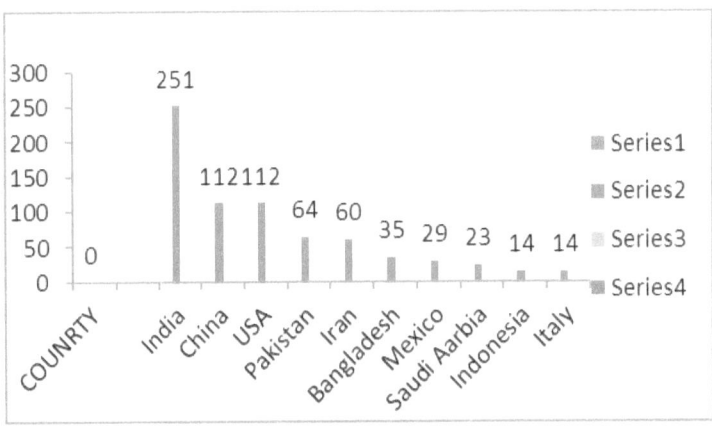

Source: Table 3.1, UN WWDR4-2012. (Figures are in Km^3/Year)

CLIMATE CHANGE AND WATER CRISIS

El Nino as Climate driver: El Nino is an unusual event that occurs once in every 3 to 8 years. The event was named by some Mexican fishermen who first noticed the changes in climate conditions during the Christmas time. During an El Nino the trade wind that blows along the equator in the pacific region from east to west relaxes or fades. This allows huge pool of warm water to flow east ward from coast of Indonesia to America. This warm water moving east ward, heats up and adds moisture to air in contact with it and changes the storm pattern and moisture delivery in hydrological cycle in America and many other parts of the world. These in effect bring sudden changes in climate and rainfall pattern approximately at an interval of five years and as an immediate consequence tropical western Pacific regions and Northern Australian regions get reduced amount of rainfall and eastern parts of South America receives more rainfall than the normal.

Greenhouse Gases (GHG) and Climate Change: The greenhouse gases such as carbon dioxide, methane, nitrous oxide including the water vapour play a vital role in earth's energy balance because they trap out going radiation to keep the surface of earth warm enough to support life. However human activities on earth raised the emission level of these greenhouse gases to such an extent that is now a concern for everyone. Before the industrial revolution, carbon dioxide level in air was 275ppm. The current atmospheric concentration of carbon dioxide is estimated at 381ppm, by far the highest level experienced since last 65,00,000 years. The result is there for everyone to realize; eleven of the twelve hottest years in instrumental records since year 1850 occurred between year 1995 and 2006. The effect of warming has other devastating implications as would there be a decline in the ability of world's oceans to remove carbon dioxide from the atmosphere, because the solubility of carbon dioxide in sea water decreases as the sea water warms. The Intergovernmental Panel on Climate Change (IPCC) Fourth Assessment Report states that the global average temperature increased by approximately 0.74°C in the hundred years up to 2006. There will multifarious effects of climate change on water demand and supply patterns.

- It will increase water shortage due to changes in the precipitation patterns and intensity around the world. Most affected regions will be in the subtropics and mid latitudes, where much of world's agriculture dependent people live. The reduced precipitation will further put pressure on ground water and a subsequent exponential drop in ground water table.
- More than 1/6th of the world's population including those in China, India, Pakistan and the western U.S live in snow melt fed river basins. Climate change and rise in global temperatures will decrease the water storage capacity from glaciers and snow cap at the sources of these river basins thereby reducing the long-term water availability in these rivers.
- It will affect the capacity of the existing water treatment plants and water distribution networks not simulated to withstand extreme weather conditions as intense floods and sustained drought in many parts of the world.
- It will increase the water demand of billions of farm animals on the planet for their hydration needs due to increase in temperature.
- It will increase quantities of water needed for industrial cooling due to increase in atmospheric and water temperature.
- It will increase the water demand for agriculture for prolonged dry periods and severe drought.
- It will create climate-forced migration of agriculture-based village population to industrial urban centres due to reduction in Agri-GDP and rural income triggering changes in water footprints.

Urbanization and changing water footprints: Nearly half the world population now lives in urban/semi urban areas. Urban population of the world is projected to grow from a level of 3.4 billion in year 2009 to 6.3 billion in year 2050. This will cause tremendous pressure on urban water supply, sanitation and drainage. More intensely affected will be the urban slum dwellers. Even by 2012, it is estimated that the world needed an almost 100 Billion ltres of water for a single day to feed the 20 most populous cities of the world.

City	Population	Water Demand (@135lpcd) in MLD	Water Demand (@250lpcd) in MLD
Tokyo, Japan	37,126,000	5012.01	9281.50
Jakarta, Indonesia	26,063,000	3518.50	6515.75
Seoul, South Korea	22,547,000	3043.84	5636.75
Delhi, India	22,242,000	3002.67	5560.50
Shanghai, China	20,860,000	2816.1	5215.00
Manila, Philippines	20,767,000	2803.5	5191.75
Karachi, Pakistan	20,711,000	2795.98	5177.75
New York, USA	20,464,000	2762.64	5116.00
Sao Paulo, Brazil	20,186,000	2725.11	5046.50
Mexico City, Mexico	19,463,000	2627.51	4865.75
Cairo, Egypt	17,816,000	2405.16	4454.00
Beijing, China	17,311,000	2336.98	4327.75
Osaka, Japan	17,011,000	2296.48	4252.75
Mumbai, India	16,910,000	2282.85	4227.50
Guangzhou, China	16,827,000	2271.65	4206.75
Moscow, Russia	15,512,000	2094.12	3878.00
Los Angeles, USA	14,900,000	2011.50	3725.00
Kolkata (Calcutta), India	14,374,000	1940.49	3593.50
Dhaka, Bangladesh	14,000,000	1890.00	3500.00
Buenos Aires, Argentina	13,639,000	1841.26	3409.75
Total		52478.42	97182.25

Source: City Populations, Largest Cities of the World. Worldatlas.com

In India the projections are, nearly 493.6 million people in year 2020, 599.40 million people in year 2030, 708.50 million people in year 2040, and 820 million people in year 2050 will live in urban areas.

Changing water foot prints in India: UN estimates the total international migrants at 214 million. As per 2011 census, the population of India was 1.21 billion. It is estimated that the total internal migrant population in India is 326 million, which is almost 28.5 percent of the total population of the country. The urban population in India was 17 percent of the total population in 1951, which is poised to rise to 42.5 percent of the total in year 2025. It is estimated that approximately 2

million people are migrating annually from rural to urban areas in India. India has about 68.9 percent rural population living in some 6.40 lacs villages. Of this, 66.2 percent rural males and 81.6 percent rural females are engaged in agriculture as cultivators or as labourers. Based on a nationwide survey conducted during the period July'2007 through Jun'2008 covering 79,091 rural households and 46,487 urban households located across the country except Leh&Kargil districts of Jammu and Kashmir and in some interior villages of Nagaland and inaccessible villages of Andaman & Nicobar Island, the National Sample Survey Office (NSSO) released its report "Migration in India, 2007-08". The report highlights the following:
(i) The major employers of rural migrants in urban areas are
   (a) Construction industries   : 41.6%
   (b) Agro based industries   : 23.6%
   (c) Manufacturing Industries : 17.0%
   (d) Transport Industries   : 16.8%
(ii) Migration of 55 percent migrant households in rural areas and 67 percent migrant households in urban areas were due to employment related reasons.
(iii) Migration pattern in the country indicates that 78 percent of rural migrants and 72 percent of urban migrants in India migrated within the same state.
(iv) The interstate migrants in India have increased from 19.6 percent in 1999-2000 to 25.6 percent in 2007-2008.

Causes of Migrations from rural to urban areas
(i) *Lack of Infrastructure*: The lack of good schools and colleges, healthcare facilities, banking and financial services, good transport and communication facilities, a suitable market in the vicinity are some of the compelling factors attributing migration of rural people to urban areas. (ii) *Employment*: Employment provides for better earning potential among rural migrants in urban areas. A large section of rural population leaves their homes and move to cities for better employment opportunities. (iii) *Climatic factors*: Intense flood and droughts cause changes in yield pattern of agricultural firms. This adversely affects earnings of landless agricultural labours in rural areas, forcing them to

move to urban centres. The mean temperature in India was projected to rise by 0.1°C to 0.3°C in Kharif and 0.3°C to 0.7°C during Rabi by 2010 and expected to rise by 0.4°C to 2.0°C in Kharif and 1.1°C to 4.5°C in Rabi by 2070. Many apple orchards in Himachal Pradesh of India have been shifted towards higher altitude regions due to rise in temperature at the middle hill regions. In hilly regions of Uttarakhand, lack of irrigation facilities and decrease in productivity of agricultural lands coupled with subdivision of land holdings owing to increased population and family size has shortened the food grain sustenance for a family to 6-8months a year and in some areas the agricultural produce is sufficient for only 1-2 months of the year. This food insecurity has caused increased migration rates among hill people to nearest urban centres to earn their livelihood.

The migration of people to urban areas also brings in dietary changes and a resultant water footprint. The amount of water consumed in production process of an agricultural or industrial product is called the virtual water or the water footprint of the product. The water footprint of some agricultural/industrial products are – 1Kg of wheat - 1654L, 1Kg of Rice-2850L, 1Kg of Maize-1937L, 1Kg of Goat meat-5187L, 1Kg of Chicken meat-7736L, a Cup of Coffee-140L, a 32MB Computer chip of 2g -3200L, a Passenger car weighing 1.1 tons – 4,00,000L, Construction of house – 60,00,000L, a Pair of Jeans – 10,850L, A Cotton shirt of 500g- 4100L. Thus, it is obvious that with urbanization and access to various utilities and a dietary shift towards meat-based products would increase the water footprint in the community as a whole. Therefore, developed countries have more water footprint per capita than those in developing or under developed countries. In Asia people consume an average of 1,400 litres of virtual water per day, while in Europe and North America people consume about 4,000litres. The rise in urban population not only results in tremendous pressure on the existing water supplies due exponential growth in water demand, but also degrades the urban ecosystem further resulting from more water abstraction and environmental pollution.

The water scarcity has resulted in many water conflicts between nations in the recent past in many water hot-spots around the world, particularly in the Euphrates, Jordan and Nile River basins, notable among them being Arab-Israel, Turkey-Syria, Egypt-Ethiopia, Israel-Palestine.

WATER HOT SPOTS AND GEO POLITICS
Jordan River Basin, Arab and Israel

The kingdom of Jordan with an approximate population of 6.30 million and a surface area of around 90,000Km² is located in the eastern Mediterranean region. It is bordered with Syria to the north, Iraq to north east, Saudi Arabia to east and south, and Israel on the west. The country receives scanty rainfall ranging between 30 to 600mm annually. More than 90 percent of the country receives less than 200mm of rainfall annually, making it one of the most arid regions of the world. Of the approximate 8.2 BCM rainfall it receives, most of it, as much as 92 percent goes back as evaporation (WWD4-2012). 37 percent of the Jordan's water supply comes from rivers. Important among them, the Jordan River and its main tributary Yarmouk River contributes to 50 percent of the surface water. A significant rise in Jordanian population has reduced its per capita availability from 906m³ in 1955 to mere 145m³ in 2008. Ground water contributes approximately 54 percent of the country's water supply. However, over abstraction has resulted in ground water deficit of 151 million m³ in 2007.

The National Water Carrier of Israel is a system of pipes and canals which carries water from Sea of Galilee to Negav Deserts. The National Water Carriers by 1990 was diverting water at the rate of 440 million cubic meters a year depriving Jordanians of all the water so diverted that used to pass through their country before 1964. Shortly after implementation of the National Water Carriers project by the Israeli authorities, in a show of solidarity with Jordan, an Arab summit held in Amman in 1964, through a resolution, decided to construct two dams on river Yarmuk. The work began on construction of two dams – Al Maquarin and Al Makhiyat on the Syrian side of the border despite stern warnings from Israel. On 5th June '1967, Israel retaliated with deep air strikes in to Syria destroying the construction works at head waters of Jordan River forcing the Arabs to call off the project. The two dams were strategically planned to divert and deprive Israel of whopping 550 million cubic meters of water a year. The Jordan and Israel reached agreement on water rights in the Jordan River in their 1994 peace treaty. However, with population of Jordan projected to reach 7.80 million by 2022, Jordan River basin with some tributaries being shared by riparian states and the kingdom Jordan itself is a water hot spot.

The Euphrates River basin, Turkey and Syria

Turkey planned to construct 22 dams and reservoirs on river Euphrates at South East Anatolian region. The works progressed accordingly on this project named as 'GAP' to reclaim 1.7mha (million hectares) of arable land. It was estimated that when the GAP project would complete, the flow of river Euphrates to Syria would reduce by 40% of its 1980 flow. The project was opposed by the Kurdish Worker's Party called PKK. Turkey contended that all waters of river Euphrates would return back to the mainstream of the river after irrigating their fields, which the Syrians opposed saying that, the same water after irrigating fields in Turkey and absorbing all the fertilizers and pesticides thereof would be much saltier and unusable even if it returned to the main stream. Syria under the president ship of Hafez-al-Assad, overtly or covertly supported the Kurdish rebels of South East Anatolia. In a bid to give a tacit warning to President Hafez-al-Assad for supporting Kurdish rebels, Turkey stopped the flow of river Euphrates in January'1990 on pretext of filling a reservoir in front of the newly built Ataturk dam. However, what Turkey did not anticipate was stopping flow of Euphrates would also cause water shortage in Iraq and would provoke the two old enemies- Syria & Iraq to join hands and stand up against Turkey. Military leaders from Syria and Iraq drew up plans to retaliate and thwart Turkey's unilateral action. Eventually on 3rd week, Turkey was compelled to allow river Euphrates to flow normally even though the filling of reservoir in Turkey was scheduled for one month.

The Nile River Basin, Egypt and Ethiopia

The Nile River consists of two branches, the White Nile originates above Lake Victoria and the Blue Nile originates in the Ethiopian highlands. Both, Blue and White Nile meet at Khartoum in Rwanda. The main Nile flows north from this confluence and passes through Egypt before draining on to the Mediterranean Sea. It is the longest river in the world with a length of approximately 6700Kms. The Nile river basin in North Eastern Africa comprising some 10 countries including Egypt, Sudan, Ethiopia, and Uganda supports a population of almost 300 million and approximately 40 percent of whole of Africa. More than 85 percent of the Nile's water come from the Blue Nile. However, majority of the Nile's flow estimated to be around 85BCM (Billion Cubic Metres) is used annually by Egypt, the last nation on the river's itinerary in to the Mediterranean Sea. River Nile has become a cultural symbol of Egypt for many centuries. River Nile sustains and supplies the fresh water needs of 60 million Egyptians living along its banks. Egypt had little or no worries for its share of fresh water from the Nile in the past. All the nations on the upstream basins were so economically marginalized that hardly any nation had resources to harness water from the Nile. Ethiopia had very little chance of getting any foreign aid to harness water from or build any dams on river Nile to irrigate their fields under the Marxist regime of Haile Mariam Mengister. However, with Marxist regime gone in 1990, a resurgent Ethiopia after a long period of civil war and famine has chosen its path to large scale economic development. With foreign aid pouring, the government is in process of implementing more than 200 small dams that will use almost 500MCM (Million Cubic Metres) of Nile water annually. Ethiopia has an estimated 3.7 million hectares of land which can be irrigated. The population of Ethiopia is nearly equal to that of Egypt but has a growth rate of 3.2% per annum compared to 2% population growth rate per annum of Egypt. Considering its future requirements to develop a large portion of its land for agriculture to sustain its growing population, additional dams are being planned to increase its irrigation and hydropower capacity. Developing Ethiopia's agricultural land and irrigating only half of its land area would reduce flow of water from river Nile to Egypt by 15%, which the Egyptians would not likely to tolerate or keep quite in future. By year 2035, the population of the Nile River basin countries is expected to increase by at least 50 percent.

Israel and Palestine

Abdel Rahman Tmini, a ground water expert from the Palestenian Hydrology Group (PHG) claims that Israelis take 80 percent of the annual flow of 615MCM (Million Cubic Metre) of water from mountain aquifers which is 'Palestinian water'. The Palestinian Hydrology Group (PHG) accuses Israeli authorities of forbidding digging of new wells in their home land since 1967. While Israel's wells are six times deeper, Palestinian wells go dry for more than five months a year. The irrigated farmland of Palestine declined from 27 percent of all agricultural land in 1967 to only 4 percent in 1990. The Israeli settlements in West Bank comprising some 1,00,000 population are given 100MCM of water, whereas the Palestinian population comprising some one million people (about ten times the Israelis), receives only 137MCM, a burning question which PHG never forgets to mention. Population of Israel is expected to rise from 4.70million (1990) to 8.00 million by year 2025. By the year 2025, because of higher birth rate, the Palestinian population is expected to reach just fewer than 7.00 million. The two communities are to share the same water resources in the year 2025, which they presently use and say are not enough and fighting over the same.

The Yellow River basin, Tibet, China and India

The yellow river basin covers an area of approximately 7, 95,000 Km^2 and supports nearly 110 million inhabitants which is approximately 8.7 percent of China's total population. Yellow River, the longest river in China traverses through Tibetan Plateau (Upper basin), the Loess Plateau (Middle basin) and the North China Plains (Lower basin). Nearly 70 percent of the basin population resides in the lower third of the basin. Many industrial cities like Xining, Yinchuan, Baotou, Zhengzhou, and Jinan came up in the basin giving rise to tremendous water demand. Since 1980, the ground water exploitation has given rise to ground water depressions in some 65 locations within the basin. The middle basin region or the Loess Plateau, covering a major part of the Yellow River basin, i.e. 6, 40,000 Km^2 is responsible for 60 percent of China's soil erosion. The river carries 1.2 billion tons of sediments each year. The silt carried by the river gets deposited along the river bed and along the banks once the river reaches the lower flood plains, where the gradient is least, before draining major portion of these sediments in to Bo Hai Sea where Yellow River terminates. As a result of decrease in flow discharge and corresponding silt deposition on the river bed, the bed of river has gone up at few places above the surrounding plains across the banks. For example, river bed is 20 meters above ground level in Xinxiang city, 13 meters above in Kaifeng city and 5 meters above in Jinan city. The silt deposits in the river bed and consequent rise in its levels has made approximately 90 million people in the basin vulnerable to floods. Notwithstanding the fact that a major flood in 1938 has affected 12.5 million people and claimed at least 8, 90,000 lives and people in the lower flood plains of the basin continue to live in fear of another major flood.

Tibet lies north of India, Nepal, Bhutan, and west of China and south of East Turkistan. It is the highest and the largest plateau on earth which stretches 2400 kilometers from east to west and 1448 kilometers from south to north. The Himalayan mountain ranges cover much of the southern boundary. The average altitude of Tibet is 3350 meters above sea level. There are more than 1000 lakes on the Tibetan plateau including world's highest Salt Lake –Namtso (Nam Co). It is estimated that Tibetan plateau has glacier cover of around 100,000 Square kilometers and store up to 12,000 cubic kilometers of fresh water. Nick named as the Third Pole after North and South Pole for its reserve of fresh water, this water tower of the world acts as the fresh water source of approximately 85 percent of Asian population and 50 percent of world population.

China annexed Tibet in 1950 with some 40,000 troops which forced the exile government in Tibet with Dalai Lama relocating and taking refuge in India. Since then Tibetan refugees all over the world are protesting against what they call Chinese aggression. China's rail tracks of 1100 kilometers from central China to Tibetan capital Lhasa constructed at a cost of approximately $ 4 billion opened in 2006. With opening of the rail communication with main land China, fierce fighting occurred in early 2008 between Tibetan protestors and Chinese police. Protestors believe that Tibet's vast resource of minerals like copper, zinc, Iron etc. will be mercilessly exploited by Chinese authorities with opening of railway lines. But many scientists and experts from other regions of the world are of the opinion that China has more strategic interest on Tibet's fresh water resources rather the minerals. "Water is seen as a strategic asset for China where ever it occurs", states Geoff Debelko, the Director of Environmental and Security Program at Woodrow Wilson International Centre for Scholars, Washington D.C. The Tibetan plateau is a home to world's most exquisite glaciers, huge alpine lakes and mighty waterfalls. Its fresh water reserves are so bountiful that it serves as the head waters of many of the Asia's largest rivers including Yellow, Yangtze, Mekong, Brahmaputra, Salween, Indus, and Sutlej among others. The Yangtze River is the third longest river in the world, after Amazon and Nile. Almost half the world population lives in the water sheds of these rivers. However, deforestation, mining, and manufacturing and other activities are producing record level of water and air pollution. "Circle of Blue", quotes report of Intergovernmental Panel on Climate Change (IPCC)'s report in May'2007 to highlight the fact that Tibet's glaciers are receding at a rate of 0.90meters annually, considered faster than any other glaciers in the world. At least 500 million people in Asia and 250 million people in China are at risk from declining glacial flows on Tibetan plateau.

A member of the Chinese Academy of Sciences, Wang Guangqian in June'2011, raised a new proposal to divert water from the upper reaches of the Brahmaputra River to country's northern province of Xinjiang. The water diversion route named "Great Western Canal" is different from the earlier diversion plan called "Western Canal" approved by the state council in December'2002. While the previous canal aims to mitigate water shortage in Qinghai, Gansu, Ningxia by diverting water from three rivers namely Tongtian, Yalong and Dadu River, the newly proposed route is expected start from the Brahmaputra River and carry water to Xinjiang along the Qinghai-Tibet Railway and Hexi Corridor – a part of Northern Silk road located in Gansu province. Due to the increase in demand along the middle and lower reaches of China's two largest rivers Yangtze and Yellow and subsequent fall in surface water reserves in these rivers, there has been a large-scale exploitation of ground water table in the cities along the lower and middle reaches of these rivers. However, the new plan as quoted by Wang Guangqianis is inspired from the "Shoutian Canal "-conceived by Chinese hydro-geologist Guo Kai in 1988. The project also called the Tsangpo Project aims to divert 200 billion cubic meters comprising nearly 33 percent of Brahmaputra River's water to Yellow River each year. The proposed dam at Shoumatan Point on the great bend on river YarlongTsangpo (as Brahmaputra known on Chinese side) would back up water that will plunge 2000 meters through 15 kilometers of tunnels and discharge through 26 of the world's biggest turbines in to a tributary rejoining the riverbed and reservoir of the storage dam at north of Medog. The water flow will be cut short from 200 kilometers to 21 kilometers. With 40 G.W of pumped storage facility, this would be world's largest hydroelectric facility. Experts believe there will be much more diversion than the originally planned 200 BCM, as the recent studies point to faster glacial melt than earlier projections of 1990 when climate change considerations were not applied. Despite repeated denial by Chinese authorities, it is believed that the Tsangpo project has already been initiated with allocation of 750M Yuan for construction of the Medog highway. The 141 kilometers Bomi-Medog high way linking lower Brahmaputra valley with Tibet's main east-west highway 318 has this unusual budget allocation when considered that Medog's population is only 10,000 and is 30 kilometers north of Arunachal Pradesh, the eastern state of India. It is believed that Medog highway would be bituminized and wide enough to carry heavy construction or military equipment. Interestingly, roughly 20 kilometers to north of Medog is the site for the proposed storage dam of Tsangpo project.

WATER SCARCE CITIES AROUND THE WORLD
Following cities around the world may go dry in future, if water resources are not managed properly-

Sao Paulo
Brazil is home to world's 2nd largest river Amazon, which is also largest in terms of water volume. Sixty three percent of the Amazon Basin in the world is located in Brazil. Sao Paulo is Brazil's largest and wealthiest city. It is the biggest city in western hemisphere with more than twenty million populations. The Cantaneira reservoirs supply water to over nine million residents of the city. The city's other two reservoirs, i.e., The Billings reservoir, which lies in the south of the city and having twenty percent more capacity than the Cantaneira reservoir and Guarapiranga are so polluted that waters of these reservoirs have become unusable.

Between, 2012-2015, a severe El Nino aggravated the drought in historically dry north east region of Brazil, but there was no excess precipitation in south east either. The rainy season during these years yielded less rainfall than usually expected during summer months. El Nino brings periods of hot spells and rise in temperature in southern parts of the country affecting many large cities in Brazil in summer. Due to successive dry spells, in 2015, the Cantaneira reservoirs were operating at twelve percent of its capacity in the month of October'2015. The water level fell so low that large parts of the reservoirs were left exposed with dried mud. For most of the year in 2015, many Sao Paulo residents experienced water cutoff for 12 hours on daily basis. At its peak, the water in the reservoirs dropped to a level having less than twenty days of water supply for the city residents. While the rich could buy water through tankers, poor consumers suffered immensely. Irony is the fact that Brazil has twelve percent of the world's fresh water resources with most of the water contained in Amazon River and rain forests in the north. During the 2014-2015 droughts, the total rain fall was fifty percent below the previous driest year on record.

Experts believe, 2012-2015, water crisis in Sao Paulo was a result of poor planning, mismanagement, and lack of maintenance of city's existing infrastructure. An independent report " Water losses: Challenges in progress of basic services and water scarcity", by G.O Associates, which used official data of Brazilian Ministry of Cities in 2013, stated that the total non-billable water in Brazil equaled 6.5 times the capacity of the Cantereira system in Sao Paulo, an equivalent of 7154 Olympic size pools a day or the entire production of the city of Sao Paulo for five years. The economic value of water wasted annually was approximately thirty nine percent of all investments in water management and infrastructure combined.

Chennai

The Chennai Metropolitan Area (CMA) including Chennai city (formerly known as Madras) and adjoining areas has a population of 8.6 million out of which 4.9 million residents reside in Chennai city. Chennai Metropolitan is the fourth largest urban conglomerate in India.

Chennai has four major reservoirs/lakes; Poondi, Cholavaram, Red hills and Chembarambakkam. These reservoirs together, supply about 200MLD of water. These reservoirs fill up by end of North East monsoon (October-November) and supply water till next monsoon season. These reservoirs over flooded due to excess rainfall causing unprecedented flood during 2015-2016.

However, due to deficit monsoon in 2019, these reservoirs dried up triggering a water crisis never experienced before. Of the total demand of 830 Million Litres per Day (MLD) of water, the Chennai Metro Water Supply and Sewerage Board (CMW&SB) could supply only 525 MLD of water. Due to consecutive monsoon deficit post flood of 2015-16 for two-three years and also in 2019, the reservoirs did not fill up. On 19th June'2019, the combined storage of all reservoirs stood at just 18 million cubic feet (mcft), which is equivalent to 0.1 percent of the total storage of 11,257 mcft.

Chennai had experienced similar drought like situation in 2000 also. After drought in 2000, rain water harvesting (RWH) installation was made mandatory in buildings in Chennai. However, rapid urbanisation, illegal encroachment of wetland areas, unregulated pumping has made the ground water table reach critical level. The number of water bodies in the city reached from sixty in 1893 to twenty-eight in 2017 and the area of water bodies shrunk from 12.6 Sqkm to just 3.2 Sqm in 2017.

Chennai now is dependent on three mega water desalination plants with a combined capacity of 180 MLD which are working overtime at eighty to ninety percent of their installed capacities to cater to the needs of water consumers having piped water supply connection.

Jakarta

Indonesia, one of the largest economies in South Asia, is home to approximately 264 million people. It is the fourth most populous country in the world. Republic of Indonesia is also the largest archipelagic state in the world having 17,508 islands. The country shares its borders with Papua New Guinea, East Timor and Malaysia. About 24 million people in Indonesia lack safe water and 38 million lack access to proper sanitation facilities.

Jakarta, the capital of Indonesia, having a population of 10 million people contributes to 17 percent of the GDP of Indonesia. Due to massive construction activities in the recent past, the green area within the city of Indonesia has reduced from 29 percent in 2007 to mere 9 percent in 2013. Lack of water harvesting structure combined with lesser green area for percolation of rainwater in to the ground has often caused floods in the city.

About 80 percent of water in the city is sourced from Jatiluhur Dam on Cita rum River and 20 percent from Cisadane and Kurkut River. However, due to limited coverage, about 30 percent of population in Jakarta remained dependent on ground water till 2017, which is becoming increasingly unusable due to contamination by septic tanks and faecal matters. It is estimated that about 50 percent of shallow wells in Jakarta are contaminated by sewage.

Beijing

Beijing is located in a semi-arid and semi humid monsoon region. It covers an area spread over 16,008 km² and has a population of 21.5 million. Beijing Municipal region received an annual average rainfall of 585mm between 1956 and 2000. About 80 percent of precipitation occurs during the month of June to September. The highest average monthly rainfall occurs in July with an average of 196.7mm and lowest average rainfall occurs in the month of December with an average precipitation of just 1.9mm.

There are five main rivers/river basins which serve as a source of water supply to Beijing region, they are- Daqing river of Daqing river basin (DRB), Yongding river of Yongding river basin (YRB), Beiyun river under Beiyun river basin (BRB), Chaobai river of Chaobai river basin (CRB) and Xun river of Jiyun river basin (JRB) and all these rivers/river basins fall under Hai river basin (HRB). Of these five rivers, Xun River, Chaobai River and Daqing River originate in Hubei province and Beiyun River is a local river in Beijing region. There are four major reservoirs, which serve as a source of water for Beijing region, ie. Guanting, Miyun, Huairou and Haizi. Of these four reservoirs, Guanting and Miyun reservoirs with a total storage capacity of 8.54 billion m³ comprises of 91 percent of total planned storage capacity for the region.

Apart from these reservoirs, there are 51, 699 constructed water wells in Beijing Municipal region of which 51,454 wells are electromechanically operated. There are two major water supply canals that supply water to Beijing city. Yongding water supply canal (YWSC) with a length of 25.40Kms carry water from 4.16 billion m^3 capacity Guanting reservoir and Jigmi water supply canal (JWSC) with a length of 105.20Kms carry water from 4.375 billion m^3 capacity Miyuna reservoir to Beijing city.

Despite such sources, there has been reduction of surface water inflow to Guanting & Miyun reservoirs. In 1980-1981, the total yearly surface water inflow of these two reservoirs reduced to an alarming 0.514 billion m^3, causing severe water scarcity in Hai river basin and Beijing region. Since the middle of 1980s, the supply of water from these two reservoirs were stopped for agricultural purpose, which on the other hand, increased the dependence on ground water and its over exploitation in the region. Owing to large discharge of waste water in the upstream of Yongding River causing serious pollution in Guanting reservoir, Beijing residents again started suffering from water scarcity in the middle of 1990s. Heavy pollution in the Guanting reservoir forced authorities to ban drinking water supply sourced from Guanting reservoir. This coupled with below average rainfall of 455mm (21.9 percent less than average annual precipitation of 585 mm during 1956 -2000) for nine consecutive years since 2000, caused severe water crisis in Beijing municipal area. The Miyun reservoir remained the only surface water source for Beijing residents for many years.

Efforts have been made to decrease water demand in Beijing region through increased water use efficiency in agriculture, educating farmers, improved irrigation management and advanced crop gene technologies. The implementation of such water saving measures has helped to reduce water demand to a tune of 4×10^9 m^3 between 2000 and 2017. But despite such water saving measures, the average annual water demand was 36.6×10^8 m^3 between 1998 and 2004 and corresponding average annual water supply remained at 23.6×10^8 m^3 for entire Beijing region with an average annual water deficit of 13×10^8 m^3. Beijing region has seen a population growth from 4.2 million in 1949 to 9.04 million in 1980 and at present 21.5 million people reside in Beijing region causing exponential growth in its water footprints due to an increased regional GDP from 10.90

billion Chinese Yuan in 1978 to 1048.80 billion Chinese Yuan in 2008, which has grown to 3.6 trillion Yuan in 2020 as per Beijing Municipal sources resulting in severe stress on water sources.

Cairo

Of the 6650Kms of river Nile, 1600Kms stretch of length passes through Egypt. The major portion of water in Nile is from Blue Nile, from Ethiopia. More than 96 percent of land mass in Egypt is desert and only about 4 percent of the country is inhabited. The average fresh water availability per person per annum in Egypt declined from 1893m^3 in 1959 to 900m^3 in 2000. The fresh water availability further reduced to 700m^3/capita/year in 2012 and by 2019 it reached a new low of 570m^3. It is expected that fresh water availability in Egypt will reduce to 500m^3/capita/year by 2025, when the country's population is projected to grow up to 98.7 million, without considering the Grand Ethiopian Renaissan Dam, which once fully operationalised will further lower fresh water availability in Egypt.

Eighty percent of Egypt's water is used for agriculture. Egypt imports almost half of its food from outside due lack of availability of water and arable land. The water scarcity is further aggravated by rise in temperature due to climate change. According to a UN WFP report, rise in temperature could drop food production to 30 percent in southern areas of the country by 2040. Some estimates suggest that, every 2 percent drop in water availability would affect one million people in Egypt, if any alteration in flow of Nile is made. More alarming is the fact that current evaporation rate of 7mm in Lake Nasser is likely to increase up to 7.3mm by 2050.

Cairo was founded by Fatimid Dynasty in 969 AD. It is located at the upstream of the delta about 250Kms from the Mediterranean Sea. Cairo city lies on east bank of the river Nile and comprises of two islands within its periphery on a total area of 453 Square Kilometres. The population of the city rose from 2.5 million in 1950 to 13.6 million in 2000 and reached 20.9 million in 2020. Over 90 percent of Cairo's drinking water demand is met from river Nile. However, river Nile receives huge quantity of untreated domestic, industrial and agricultural waste water in upstream of the Cairo city. Between Aswan Dam and Cairo, waste water from 43 towns with population exceeding 50 thousand and some 1500 villages dump their wastes into river Nile. It is estimated that each year, about 2.3 billion cubic metres of drainage water from irrigated fields loaded with fertilizers, chemicals, pesticides and organic materials, and more than 125 million cubic metres of industrial waste water find their way into river Nile upstream of Cairo city making every common citizen in the city vulnerable to various water borne diseases. With population rising over 21 million by 2025, the city of Cairo depending on its only source of fresh water in river Nile waits for a water crisis that looms large in future.

Delhi
Delhi is surrounded by populous cities like Ghaziabad, Faridabad, Gurugram and Noida. A large number of floating populations visit the city on daily basis from these cities due to its functional linkages. In fact, population projection by United Nation predicts Delhi's population to outnumber Tokyo region by 2028, with population of this urban agglomeration reaching 37.2 million in the projected year. Ninety percent of Delhi's raw water intake is sourced from surface water, i.e. rivers and canals and rest nine percent from ground water. As per data provided by Economic Survey of Delhi 2019-20, Planning department, river Yamuna provides 1438 million litres, Ganga River provides 946 million litres and Bhakra storage provides 837 million litres of raw water for Delhi residents on daily basis. Another 325 million litres per day of raw water is taken from ground water sources to meet daily water needs of Delhi city, making total raw water availability for the region as 3546 MLD (Million Litres per Day).

As per 2011 Census of India, the population of Delhi city was 16.79 million, with 97.5 percent classified as urban and 2.5 percent as rural population. The population of the city is expected to reach 23 million in 2021. As per Delhi Jal Board, the city requires 172 lpcd (Litres Per Capita per Day) of treated water for its domestic needs and another 102 lpcd to meet non-domestic demands such as industries, commercial establishments, hotels; fire stations etc. making total daily per capita demand as 274 lpcd. Thus, daily water demand of the city with a population of 23 million in 2021 would be 6302 million litres, giving a deficit of supply to the tune of 2756 million litres of water on daily basis. Apart from this huge demand and supply gap of water availability for Delhi city, during summer neighbouring states like Haryana, Rajasthan sometimes curtail water flow to Delhi region due to demand in their own states. This further causes the severity of the problem. Many localities in Delhi, depending on ground water sources have seen ground water depletion of 25-30 mtrs during the last decade due over abstraction of ground water. The leakages in water distribution pipe lines, water theft through illegal water connections, faulty water meters etc has resulted in a non-revenue water (NRW) of 40 percent for Delhi Jal Board's water supply. This has resulted in a budget deficit of Rs. 1.77 billion (2018-19). In many farm houses within the city boundaries, ground water is being used for watering lawns, washing vehicles etc. To reduce ground water depletion, installation of rain water harvesting systems in apartments and private buildings with more than 100 square metres were made mandatory for new constructions since 2001, however no steps have been taken to implement such regulations by concerned authorities. As a result, ground water depletion is continuing uninterrupted due to over abstraction by about 25 percent, resulting in increase in ground water salinity, and higher than prescribed limit of fluoride, nitrate and arsenic in ground water in the region.

Mexico City

Mexico City situated at an elevation of 2250 miles above sea level covers an area of 1485 Sq.Km. The population of the city increased from 3.36 million in1950 to 8.83 million in 1970 and 18.45 million in 2000 to 21.78 million in 2020, thus showing a receding trend in population growth during the first two decades of the 21st century. The average minimum rainfall is 1mm & 2mm in the month of November & December respectively and the city receives average maximum rainfall of 61mm, 70mm & 55mm during the month of June, July and August respectively.

The city was founded by Aztecs in year 1325 on an island surrounded by network of lakes. These lakes provided a line of defense to the Aztecs against its enemies. Later, Spaniards replaced the dikes and canals around this Aztecs capital with streets and squares; drained the lakes and cleared the forests making the city vulnerable to floods owing to ecological imbalance in the aftermath. By the time Mexico got its independence, most of the fresh water was gone. The city itself has gone tremendous transformation during the second half of last century, with its size growing from 50 Sq.kms in 1950 to 1485 Sq.Kms. Due to continuous floods during rainy seasons, a major tunnel project costing more than 2 million dollars, to drain out flood waters, was taken up. The system helped to move water out of the city quickly without giving any appreciable time for ground water recharge from flood waters. The city's water distribution pipes are more than 50 years old. As per a government study in 2010, the distribution loss in water pipe lines was around 40 percent. There is disparity in daily supply of water to city residents, with people in rich neighbourhood getting more than 1000 litres of water and poor residential colonies receiving less than 200 litres. Residents in many neighbourhoods depend on water tankers for their sustenance, which supply water once or twice in a week. These water tankers called 'Pipa', charge hefty amount for their supply.

Agriculture accounts for 77 percent; Industry uses 10 percent and 13 percent of all water is used for domestic purpose in Mexico. With increase in population, the residents of North, North West and Central regions of Mexico uses an average of 300 litres of water per day, which is more than their US counterparts. These regions have 77 percent of population and responsible for 84 percent of GDP of Mexico. Massive rise in population coupled with only 28 percent of runoff water supply has given rise to acute water scarcity in the region. Over extraction of ground water has seen subsidence in many parts of the city with sewer pipes cracked, roads broken in places, and most notable is tilting of the iconic cathedral in Mexico City's historical centre. The city now imports as much as 40 percent of its drinking water from distant sources, which is pumped to a height of more than 1000 metres through a series of pumps which consumes as much energy as the entire metropolis of Puebla, a Mexican state capital.

An independent study predicts 10 percent of Mexicans aged between 15 to 65 years would eventually emigrate to north as a result of rise in temperature, intense draught and floods due to climate change.

Cape Town

The Cape Town region has a Mediterranean type of climate. The summers are dry and warm and rainfall occurs in winter. The water usages are – 70 percent on urban and 30 percent for agriculture purposes. The water is sourced from six major dams located in the mountainous region. These dams fill during rains in cooler months of May through August. The water levels in the dams reside during dry summer months of November to April, when most water is used in urban region and in irrigating crop fields.

Following low winter rainfall in 2000-2001 and again in 2003-2004, the construction of Berg River Dam and Supplement Scheme was taken up. With completion of this dam and supplement scheme in 2009, the storage capacity of the dams supplying water to Cap Town city increased by 17 percent from 768 million cubic metres to 898 million cubic metres. The city had an efficient water management system. The water loss management and conservation efforts in the city helped it to restrict its leakages in water distribution systems below 15 percent which was lower than 35 percent national average in 2015. These efforts of the city have attracted many international awards in Water and Demand Management during the same year. However, things started to fall apart, when storage capacity of all six dams, i.e. Berg River Dam, Steen bras lower Dam, Steen bras upper Dam, Theewaterskloof Dam, Voelvlei Dam, Wemmershock Dam together decreased from 646.14 billion litres in May'2014 to 450.43 billion litres in May'2015, a marked decrease of water storage in the dams from 71.9 percent to 50.1 percent with water level reaching half the storage capacity of these dams. Again, owing to scanty rainfall, the combined water quantity in the dams plummeted to 279.95 billion litres in May'2016, which was 31.2 percent of total capacity. In November'2016, the Department of Water & Sanitation imposed various restrictions for urban and agricultural use of water. Due to continued low rainfall, the water levels in all six dams accounted for combined storage of 190.3 billion litres on 15th May'2017, which was merely 21.2 percent of the combined capacity of all the dams supplying water to Cape Town city. On 1st June'2017, Level 4 restrictions were imposed on water consumption, limiting 100 litres of water per person per day, a limit lowest since 1933. In January'2018, then Mayor of Cape

Town city, Patricia de Lille announced that 'Day Zero' restrictions would be imposed, if situation do not improve and municipal water supplies would be stopped and city residents would be required to collect water through 149 water collection points where each consumer would get 25 litres of water. The 'Day Zero' was initially projected for 22 April'2018. By February'2018, Level 6B restrictions were imposed limiting usage of water to just 50 litres per person per day in city of Cape Town.

However, a good winter rainfall in 2018 saved the city from an impending disaster when the dam levels reached 76 percent at its peak. The water storage capacity in the dams supplying water for city of Cape Town increased 17 percent during 1995 to 2015; but the city population has increased from 2.4 million to 4.1 million during the same period, an increase of 71 percent over a period of 20 years. With climate change taking its toll, and a prediction of 0.25 degrees rise in temperature for the city by 2030, Cape Town city may face another crisis in future.

WATER CONFLICT CHRONOLOGIES

It is not always that the conflicts over water are between two or three riparian countries. Dr. Peter H. Gleik in his "Water Conflict Chronology" brings to fore following major events where infighting took place within the boundaries of a nation or state too:

In 1999, fighting erupted between two villages over a spring located near Ta'iz in Yemen. The village of Al-Marzuh where the spring is located believed it had the exclusive rights to the water from the spring because of its location, which the neighbouring village of Quradah vehemently resisted. The village of Quradah claimed that their rights were conferred by a 50-year-old court verdict. Subsequently intense fights erupted between the two villages resulting in death of six villagers and injury to 60 others. 700 soldiers were sent to quell the fighting which was later, intervened by the President Ali-Abdullah Saleh summoning both the village Sheikhs and sorting the problem by distributing water in two halves.

In 2005, fighting erupted between Kikuya and Masai communities in north western Kenya. Tension started when a local Kikuya politician diverted river water to irrigate his farms in the upstream side of a grazing field being used by the Masai herdsmen. The infighting between the two warring factions led to death of twenty people in January alone and by July'2005, the casualty touched ninety, principally in the rural centre of Turbi.

Some other notable events as listed by Dr. Peter H. Gleick, Pacific Institute for Studies in Development, Environment and Security, also include:

Year	Parties Involved	Description
2500BC	Lagash, Umma	Lagash – Umma Border Dispute – Urlama, The King of Lagash from 2450 to 2400B.C. diverted water from this region to boundary canals, and dried ditches to deprive Umma of water. His son later cut off water supply to Girsu, a city in Umma.
720-705BC	Assyria, Armenia	After defeating the Halidians of Armenia, Sargon II of Assyria destroyed their intricate irrigation network and flooded their land.
705-682BC	Sennacherib, Babylon	In quelling rebellious Assyrians in 695 B.C., Sennacherib razed Babylon and diverted one of the principal irrigation canals.
669-626BC	Assyria, Arabia, Elam	In campaigns against both Arabia and Elam in 645 B.C., Assurbanipal, son of Esarhaddon, dried up wells to deprive Elamite troops. He also guarded wells from Arabian fugitives in an earlier Arabian war. On his return from victorious battle against Elam, Assurbanipal flooded the city of Sapibel, considered an ally of Elam.
539BC	Babylon	According to Herodotus, Cyrus invaded Babylon by diverting the Euphrates above the city and marching the troops along the dry riverbed.
1573-74	Holland, Spain	In 1573 at the beginning of the eight-year war against Spain, the Dutch flooded the land to break the siege of Spanish troops on the town Alkmaar. The same defense was used to protect Leiden in 1574. This strategy later became known as the Dutch Water Line and was used frequently for defense in later years.
1777	United States	During the War for Independence, The British and Hessians attacked New York water works.
1841	Canada	A reservoir in Ops Township, Upper Canada (now Ontario) was destroyed by neighbours considering the same, a health hazard.
1850	United States	Local residents unhappy over its effect on water levels, attacked a New Hampshire dam that impounded water for factories downstream.
1898	Egypt, France, Britain	A French expedition's attempt to control the headwaters of White Nile, almost led to a military confrontation between Britain and French in 1898.

Year	Parties Involved	Description
1935	California, Arizona	Arizona called out the National Guard and militia units to the border with California to protest the construction of Parker Dam and diversions from the Colorado River; dispute ultimately was settled in court.
1943	Britain, Germany	British Royal Air Force bombed dams on the Mohne, Sorpe and Elder Rivers. Mohne Dam breach killed 1200 people, destroyed all downstream dams for 50Km. The flood that occurred after breaking the Elder dam reached a peak discharge of 8500m^3/s, which is nine times higher than the highest flood observed. Sixty-eight people were killed.
1948	Arab, Israel	Arab forces cut off West Jerusalem's water supply in Arab Israel war.
1960s	North Vietnam, United States	Irrigation water supply systems in North Vietnam were extensively bombed during Vietnam War. Six hundred sixty one sections of dikes damaged or destroyed.
1967	Israel, Syria	Israel destroyed the Arab diversion works on the Jordan River headwaters.
1970s	Argentina, Brazil, Paraguay.	Brazil and Paraguay announced plans to construct a dam at Itaipu on the Parana River, causing Argentina concerned about downstream environmental repercussions and the efficacy of their own planned dam project downstream. Argentina demanded to be consulted during planning of Itaipu but Brazil refused. An agreement was reached in 1979 that provided for the construction of both Brazil and Paraguay's dam at Itaipu and Argentina's Yacyreta dam.
1978	Sudan	Two students died during demonstrations in Juba, Sudan against the construction of the Jonglei Canal.
1982	Guatemala	One hundred seventy-seven civilians were killed in Rio Negro over opposition to Chixoy hydroelectric dam.
1990	South Africa	Following protests over miserable sanitation and living conditions, Pro-apartheid council cut off water to the Wesselton township of fifty thousand black Africans.
1991	Iraq, Kuwait, US	Baghdad's modern water supply and sanitation systems were intentionally and unintentionally damaged by Allied coalition during the Gulf war in 1991. Four of seven major pumping stations were destroyed, as were 31 municipal water and sewerage facilities – 20 in Baghdad, resulting in sewage pouring in to Tigris.
1998-2000	Eritrea, Ethiopia	Water pumping plants and pipelines in the border town of Adi Quala was destroyed during the civil war between Eritrea and Ethiopia.
1999	Bangladesh	Protest led by former Prime Minister Begum Khaleda Zia to highlight deterioration of public services like water and power shortages, turned violent; 50 people hurt.

Contd...

Year	Parties Involved	Description
1999	Puerto Rico, U.S.	Following chronic water shortages in the neighbouring towns, Protestors blocked water intake to Roosevelt Roads Navy Base protesting U.S. military presence and Navy's use of the Blanco River.
2000	Hazarajat, Afghanistan	Due to extreme drought and subsequent depletion of local resources, violent conflicts broke out over water resources in the villages BurnaLegan and TainaLegan, and other parts of the region,
2002	Kashmir	Two people were killed and 25 others injured in Kashmir when police fired at a group of villagers clashing over water sharing from an irrigation stream near Garend village.
2004	Mexico	Two Mexican farmers argued for years over water rights to a small spring used to irrigate a small corn plot near the town of Pihuamo and subsequently shot each other dead in March'2004.
2004	China	Tens of thousands of farmers staged sit-in demonstrations against the construction of the Pubugou dam on Dadu River in Sichuan Province. Riots police were deployed to quell the unrest in which one person was killed. Witnesses also report the deaths of a number of residents.
2004	South Africa	Poor delivery of water and sanitation services in Phumelala Township led to several months of protests, including some severe injuries and damage to municipal properties.
2004	India	Four people were killed in October'2004 and more than 30 people injured in November same year during protests by farmers over allocations of water from Indira Gandhi Irrigation Canal in Sriganganagar district, which borders Pakistan. A curfew was subsequently imposed in towns of Gharsana, Raola and Anoopgarh.

WATER CONFLICT INVOLVING PUBLIC VS PRIVATE ENTITY

The Coca Cola Virudha Samity Vs the Hindustan Coca Cola Beverages:

Plachimada is a village under Perumatty panchayat in Palakad district of Kerala state in India. The area around Plachimada village is predominantly agriculture based. Paddy being the major crop and in some places sugar cane is also grown. The main sources of water for villagers living in Plachimada are the canals which flow from Malampuzha Reservoir. The water from these canals irrigates thousands of acres of land around these areas. But even than water has been in short supply despite of existing canals. In 1999, the Hindustan Coca Cola Beverages was issued licence to produce Coca Cola brand of beverages including mineral water. Under this licensed agreement, the company produced 200 truckloads of beverages containing 550-600 cases of soft drinks on daily basis. In process company extracted 6,00,000 litres of ground water from 6 bore wells every day in its factory on 38 acres plot in Plachimada. As per one Audit Report company produced around 600 cases of soft drinks comprising of 24 bottles of 300 ml capacity each, every day.

Plachimada, being located in a rain shadow area, the recharge water received from rainfall is lower than rate at which water was extracted from these bore wells. As a result, the water level dropped unusually turning the village and adjoining area, water stressed within a short period.

On April, 7th, 2003, the Perumati village Panchayat took a decision to cancel the licence agreement granted to the company. This decision was challenged by the Hindustan Coca Cola Beverages in Kerala High Court. On 16th December, 2003, the Kerala High Court directed the company to use only that quantity of water which was available to the owner of 34 acres of land. However, continued protest and litigation helped the people of Plachimada to shut down the factory in March'2004.

On 15th, January 2005, the Coca Virudha Janakeeya Samara Samity (Anti Coca Cola People's Struggle Committee) and the Plachimada Solidarity Committee spearheaded the struggle against the company and demanded permanent closure of the factory.

In March'2005, the Perumatty Panchayat refused to renew the license to the Cola Company forcing the company to shut down its operations. On 7th April, same year, the Coca Cola company appealed to High Court of Kerala, which gave a further ruling which allowed the company to extract 5,00,000 litres of water.

On 19th August'2005, Kerala State Pollution Control Board directed Coca Cola Company to stop making all produce with urgent effect because of high level of Cadmium around the plant. (As per an agreement with Kerala State Pollution Control Board (KSPCB), up to 1.5million litres of water were drawn from 6-bore wells situated inside the factory compound. The permit granted Coca Cola, the right to extract ground water at rate of 3.8 litres of water needed to produce 1(one) litre of Coca Cola.).

In September'2009, a thirteen-member high power committee was formed to assess the damage caused by the Coca Cola factory. The committee, after holding four hearings and visiting factory location and affected areas, came to conclusion that Cola Company was liable for damages caused to village economy and its environment, agriculture losses, health problems and pollution of water resources. The committee recommended creation of a "Plachimada Claims Tribunal" by state legislature to claim compensation of Rs. 216.26 Crores from the Coca Cola Company. On 16th February'2011, the state cabinet approved a draft bill, which was passed shortly thereafter in legislative assembly, to form a tribunal for securing compensation worth Rs. 216.26 Crores from the company.

World Water Council

The World Water Council is the International Water Policy Think Tank, founded in 1996. It brings together 294 members from about 40 countries in a unique network comprising of public institutions, private sector firms, United Nations organizations and non-governmental organizations. Its mission is to promote awareness of critical water issues at all levels, including the highest decision-making level, to facilitate efficient conservation, protection, development, planning, management and use of water in all dimensions on an environmentally sustainable basis for the benefit of all life on earth.

In August' 1998, the World Water Council convened the "World commission on water for 21st century" in order to look at the challenges and create a shared vision for water, life and environment. The Commission in its report stated:
- Only 10 percent of river Nile reaches the Mediterranean Sea and what does flow in to sea is polluted with agricultural, industrial and municipal waste.
- China's yellow river dried up for more than half the year in 1997.
- The flow of Russia's Amu Darya & Syr Darya Rivers in to Aral Sea in central Asia has been reduced by 75 percent for use in cotton irrigation. This has caused the sea levels to recede 53 feet between 1962 and 1994.

- Because of its extensive use in irrigation of 3.7 million farm acres, little is left in the USA's Colorado River to protect the ecosystem downstream, which has turned in to salty, lifeless marshes.
- Contamination of rivers and river basins around the world has displaced an estimated 25 million environmental refugees in 1998, which exceeded for the first time the estimated 21 million refugees displaced due to warfare.

Report also stated that more than half the world's rivers were going dry or were polluted. Of the 500 major rivers in the world, Amazon River in South America and Congo River in Sub Saharan Africa are the healthiest because they have few industrial centres near their banks.

Disappearing Aral Sea

Aral Sea, a basin shared by Afghanistan, Iran and five other countries of former Soviet Union was once the world's fourth largest lake fed by two of the prominent rivers-Amu Darya and Syr Darya. In 1963, the surface of the Aral Sea measured 66,100 Km^2 with an average depth of 16 meters and a maximum depth of 68 meters. In the early sixties, the planners of erstwhile Soviet Union decided to irrigate the central plains of Asia and deserts of Uzbek and Kazakh with water diverted from Amu and Syr Darya Rivers in order to grow cotton. In fact, with exploitation of these rivers and use of fertilizers and pesticides, between 1940 and 1980, Soviet Union remained the second largest producer of cotton in the world. However, as a result of these diversions in Amu and Syr Darya Rivers, by 1987, 27,000Km^2 of former sea bottom of Aral Sea had become dry land. Aral Sea had lost 60 percent of its original volume with decline in its depth as much as by 14 meters. Satellite pictures reveal that in 1989-1990, the Aral Sea has separated in to two parts; the "Large or South Aral and the "Small or North Aral". Between November' 2000 and June'2001, a deserted island in the sea reportedly used by former Soviets to test their biological weapons joined the mainland, opening a flood gate for further pollution. Today, about 2,00,000 tons of salt and sand are carried by the wind from the Aral Sea region every day and dumped within a 300 kms radius, destroying pastures and decreasing available agricultural land each passing day.

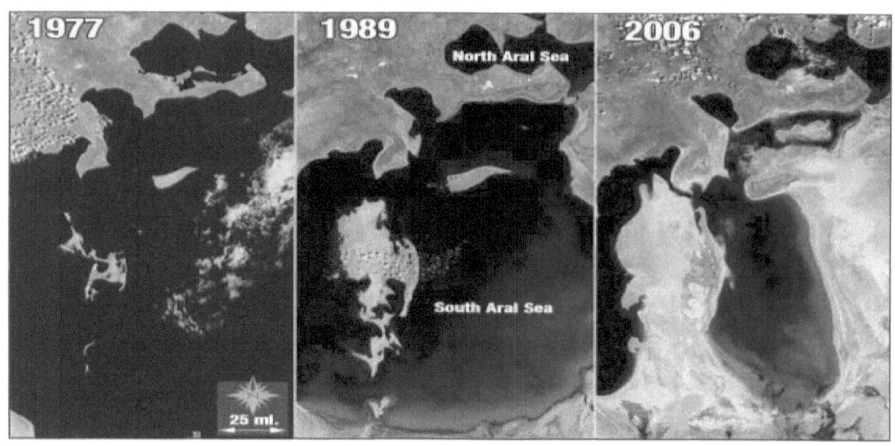

Disappearing Aral Sea. (NASA photographs of Aral Sea at respective years.)

Chapter 3

MANAGEMENT

Few suggestions to mitigate water scarcity in future would be-

Encouraging traditional water harvesting methods: Traditional rain water harvesting techniques like bamboo drip irrigation of Meghalaya, terrace cultivation in most north eastern states, ahar-pynes in Bihar, tankas/kunds and khadins in Rajasthan, zings in Kashmir, traditional Erys in Tamil Nadu should be encouraged by the states. These age-old wisdoms of collecting and conserving rain water are showing fulfilling results in many states now

Encouraging Agro-forestry: Agro-forestry is a blend of agriculture and forestry, where the crop area is 80 percent or more and tree area - 20 percent or less. The trees planted are usually nitrogen fixing trees like anjan, acacia, shisham, siris, casuriana, khejri etc. Nitrogen fixed by these trees besides enhancing the soil fertility are also utilized by the associated crops. Such system of land management also enhances the uptake of carbon dioxide and has the potential to remove significant amount of carbon dioxide in the atmosphere. Research done with Tephrosialvogelii & Gliricidiasepium over a period of ten years in Malawi, Africa showed that Maize yields averaged 3.7 tonnes per hectare compared to 1 tonne per hectare in plots without nitrogen fixing trees or mineral fertilizers.

Saving water through virtual water trade balance: The water scarce nations could import water intensive crops from other countries and thus save water and utilize the cultivated land for less water intensive crops having lower water footprints. Similarly, within the nation itself regional balance of water could be achieved by relocating water intensive crops from water scarce regions. But maintaining virtual water balance could also give rise to unemployment and other local problems. India has exported 228.61Gm^3 of virtual water with an average of 45.72Gm^3/year and imported 358.27Gm^3 of virtual water with an average of 71.65Gm^3/year during the period 2001 -2005. Thus, India saved around 129.66Gm^3 in virtual water during the same period through its export/import of various agricultural and industrial commodities.

Conservation through water harvesting legislations: Nearly 50 percent of the global population is now living in cities or urban areas, and people living in these urban/semi urban areas are also responsible for approximately 70 percent of the greenhouse gas emissions around the world. Many governments like Germany and the USA are now putting rain water harvesting legislations in their building bye laws and provide incentives for water conservations. Residents of San Antonio in Texas can apply for a tax rebate up to 50 percent for their rain water harvesting projects.

Management by augmenting supply through Rain Water Harvesting: Noted Slovakian hydrologist Michael Kravcik and his NGO "People and Water", studied the effect of urbanization in his own country Slovakia, which is a small nation in central Europe. The country has fast developed in to an urbanized state within a short period of time. With all developmental works, the state has lost most of its forest covers and wet lands; most of grass cover has given way to paved surfaces. Kravcik believes every drop of rain water has a "Right of domicile" where it falls. But as soon as the forest is denuded or the grass cover is removed to pave way for concrete buildings or for that matter wet lands are destroyed due to developmental works, the right of domicile of a rain drop is lost. The rain water that was supposed to percolate in to ground or reach the inland ponds or lakes and mingle with the natural environment now falls on the paved surface and passes as runoff and reaches the sea and converts to salt water. In short, every time a surface is paved, there is an increment to fresh water conversion in to salty sea water. Kravcik and his team actually studied this effect of urbanization resulting in additional roofing, residential paved surfaces, paved high ways, parking etc. on the reduction of water supplies. Every year about 250 million cubic meters of fresh water disappears in Slovakia- an equivalent of 1 percent of all water in Slovakia's water sheds. Since World War II, Slovakia has lost 35 percent of its annual precipitation.

RAIN WATER HARVESTING SYSTEMS
Roof Top Rain Water Harvesting Systems

A roof top rain water harvesting system should have the following essential features-

- The roof for collecting rain water should be made of non-toxic materials.
- The roof surface should be smooth, hard and dense so that it is easier to clean and less likely to get damaged or for rain water to carry with it the roof fibre element.
- It is not advisable to paint the roof as most of the common paints carry toxic substances which may peel off due to seasonal variations of hot and cold conditions of the surface and under influence of heavy showers of long durations.
- The dry leaves from overhang branches of trees tend to accumulate on the roof surface. Therefore, it is not advisable to have any overhanging trees near the roof.
- To prevent leaves etc. from reaching the filters or the first flush devices, all gutter ends should be screened with wire mesh.
- Nesting of birds on the roof should be prevented. A whole range of pathogens has been detected in the rain water collected from the roof tops which include Salmonella, Campylobacter, Giardia, and Cryptosporidium. The likely sources of these pathogens are the faecal materials deposited by birds, frogs, rodents, dead animals, insects etc. either in the gutters or in the water tank itself.
- The storage tanks should have a tight roof cover which excludes light, and a flushing pipe at the bottom of the tank.
- The storage tanks should be provided with a reliable sanitary extraction device to clean and extract water from time to time to prevent contamination of water in the tank.
- Base of the storage tanks installed at ground level should be raised so that there are no chances of waste water flowing in to the tank and contaminate stored water inside.
- Unless reliable, water from other sources should not be allowed to collect in to the rain storage tanks through any pipe connections or through manhole covers.

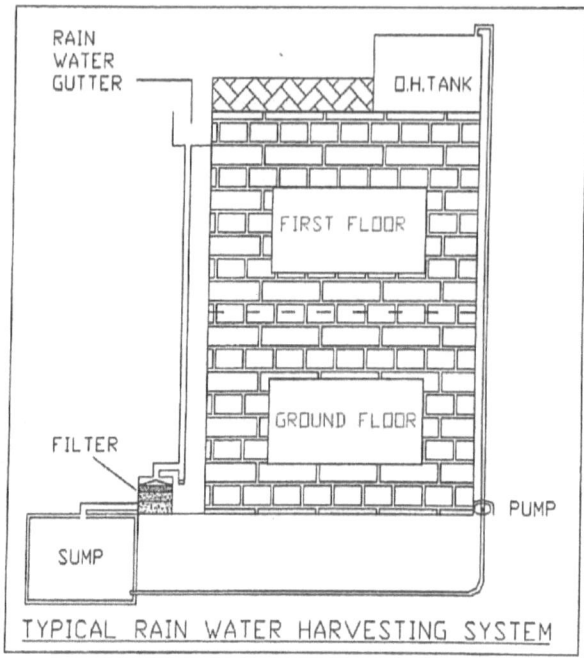

(ACAD representation: Author)

Components of Rain Water Harvesting Systems

Catchment: It is the surface, which directly receives rainwater and provides water to the system. It may be paved surface like terrace of a building or the courtyard of the building or it may be an unpaved surface within the compound like a lawn or open ground. Thus, an RCC roof or a CGI sheet roof also acts as a catchment of a rainwater harvesting system.

Coarse Mesh: It is a mesh of screening materials to prevent debris from entering the rain water pipes. It is placed at the inlet to the pipes.

Gutters: They are the channels provided all around the roof surface to collect rain water. Basically, they may be of two shapes – (a) rectangular (b) Semicircular. Gutters are made of usually (i) Plain GI sheets (20/22 gauge) folded to required shapes, PVC pipes cut in half, Straight bamboo or beetle trunk cut in half. The size of the gutter should be able to take discharge from the rainfall of highest intensity.

Conduits: These are the pipelines that carry rainwater from the rooftop to storage tanks or to the recharge pit. These pipes may be PVC, or GI depending upon the choice of the users. In general PVC pipes are preferred due lower costs and easy handling.

National Building code (NBC) has given some diameter of pipes to be used with different intensities of rainfall and different roof areas for ready reference of the users.

First Flush: Normally before the rainy season starts, the roof surface, if not periodically cleaned, will contain films of dust and dirt, leaves, bird droppings and other pollutants which when rains if reaches the water tank will pollute the water and make it unsafe to drink or use for any other purpose. The dirt and debris that may flow with the rain water through the gutters and pipes may also clog the rain water filter rendering it unfit for use. Therefore, a device is used to divert the first shower of rain that contains dirt, debris, bird droppings etc. to an outfall so that it does not reach the rain water filter. This device is called the first flush.

There are two types of first-flush devices, they are-

- *Floating ball first-flush*:- This type of first-flush device contains a light hollow floating ball (preferably plastic) inside a vertical casing pipe which is connected to the rain water pipe laid in horizontal direction above. The rain water pipe is connected to the pipe housing the floating ball through a reducing 'Tee'. The vertical casing pipe has a slot of 1(one) inch diameter to clear way the first flush of rain. When the first water from the rain moves through the rain water pipe, it falls in to the vertical casing pipe holding the floating ball. When the water level inside the vertical casing pipe rises the floating ball gets lifted along and finally blocks the flow of first flush of dirty rain water at the mouth of reducing tee. Thus, the comparatively clear water now moves through the horizontal main pipe and reaches the filter. The first flush of dirty water from the casing pipe is then drained out through the slot at the bottom of the vertical casing pipe. (Fig)

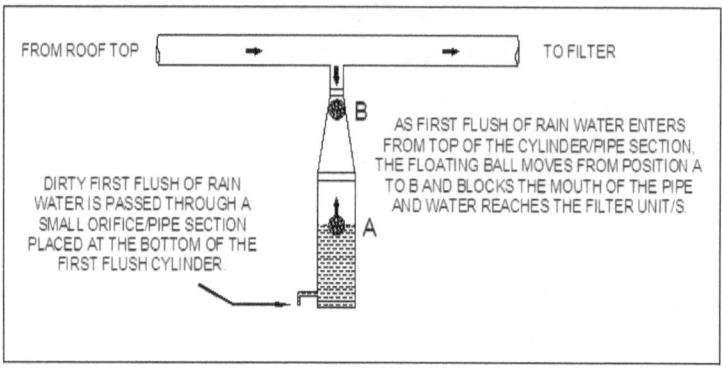

(ACAD representation: Author)

- *Tipping gutter first-flush device:-* This device contains a gutter which is connected through a pulley to a container to receive the first-flush of rain water. When the first rainfall falls on the gutter, it reaches the container connected to it. As the volume of rain water so collected in the container increases, the weight of the container increases and moves down, thereby lifting the gutter through the string connected via pulley thereby stopping further the entry of first rain water in to the container and the comparatively cleaner water reaches the water tank.

Guide lines for first flush devices
- The first-flush devices should be able to remove dirty rain water from the first rainfall at the rate of 10gallons per 1000Sqft of roof area or the catchment. i.e. about 37.8 litres for each 92 Sqm of roof area.
- If the harvesting rain water is done in an area containing lots of dusts or other pollutants also with poor air quality, first-flush device has to be sized accordingly.
- The first-flush device should have an easy access for maintenance otherwise the same is likely to be futile and disabled.

Filter:- After the first flush of rain is passed out, it is allowed to pass through a filter. The various types of filters, presently being used are:-
(i) *Charcoal filter:* A Charcoal filter can be made out of a drum or an earthen pot. The filter media consists of gravel, sand & charcoal, which are easily available.

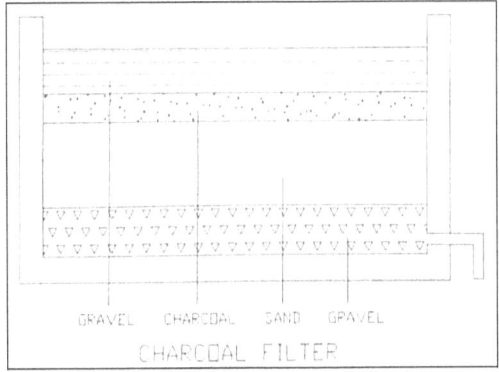

(ii) *Sand filter:* The primary filtering media in Sand filter is Sand, which is commonly available. These filters can effectively reduce turbidity. The top layer consists of coarse sand followed by 5-10cm layer of gravel and then the bottom layer comprises of 5-25cm layer of gravel and boulders.

(iii) *Dewas filter:* In Rajasthan, India, most of the residences in Dewas district have a well in the premises. During the severe drought in the year 2000 that swept Rajasthan along with other parts of the country, most of these wells dried up due to ground water depletion. Mr. Mohan Rao, I.A.S, the District Collector of Dewas during that period initiated a process of recharging these wells through rainwater harvesting. He has designed a filter that can effectively filter the rainwater. This filter is now popularly called the Dewas filter. This essentially has a PVC pipe of length 1.20mtr & 140mm dia which contains three layers of filtering media – first layer contains 2-6mm pebbles, second layer contains 6-12mm pebbles and the third 12-20mm pebbles. A mesh is provided at the out flow of the filter. The cost of this filter was about Rs. 600/-(2002).

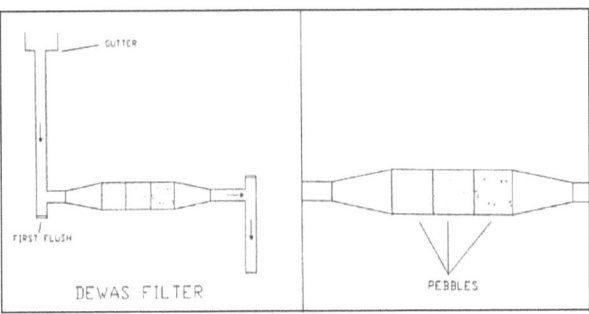

(iv) *Jeyakumar Filter*: These are the filters for roof tops with larger areas. This has been designed by Mr. R Jeyakumar. To increase the area of filtration, the sand, gravel & pebbles are placed in three concentric chambers. The outer chamber is filled with sand, middle chamber is filled with gravel and the innermost chamber is filled with pebbles. This way the area of filtration through the sand layer increases. The water enters the outer chamber and flows in to the centre core from where it is collected in the sump. And the rainwater so collected is used after necessary disinfection.

(v) *Varun Filters*: This is one of the rainwater filters that is now commercially available. According to its inventor Mr. S. Vishwanath, a resident of Bangalore, One Varun has the filtering capacity to handle 50mm/hr. intensity rainfall from 50Sqm of roof area and is standardized to meet the requirement depending on rain fall intensity & roof area. i.e. 100Sqm of roof area would need 2nos of such units. It is essentially made of High-Density Poly-Ethylene (HDPE) drums of 90litres capacity each, with its lid turned over and holes punched in it. This acts as the first sieve which keeps out large leaves, twigs etc. from entering the filter. Rainwater coming out from the lid passes through three layers of sponge and a 150mm layer of coarse sand. The sand does not need cleaning at all. Simply the first layer of sponge has to be cleaned from time to time which can be easily removed, cleaned and placed it back in the filter.

Storage tanks: The Storage tanks or the Cistern is one of the important components of a rainwater harvesting system. This is because the type and capacities of such tanks used for domestic consumption affect investment decisions in such systems. The volume of the storage tanks depends mainly on following factors:

(i) Number of persons in the household.
(ii) Per capita water requirement.
(iii) Average annual rainfall.
(iv) Size of catchment or the roof area.

There are three practical approaches while designing the capacity & size of Storage Tank.
(i) Matching capacity of the tank to the area of the roof.
(ii) Matching the capacity of the tank to the quantity of water required by its users.
(iii) Choosing a tank size that is appropriate in terms of cost, resources & construction methods.

A study in March-April'1980 on rainwater collected from roof on an office building in central Tokyo found that 0.5-1.0mm of rainfall was needed to wash the surface of the roof. Another research done in August-September'1986, also in central Tokyo, measuring the degree of change in water quality in rainwater collected from a roof after a period of 11 days of no rain, showed that water quality is generally stable after 1.50mm to 2.00mm of rain. In other words, even after a dry spell, only about the first 1.5 mm of rainwater is affected by pollutants accumulated on collecting surface. Rainwater users can reduce their risk of disease from contaminated rainwater consumption by regular maintenance and using a well-designed system. A range of pathogens has been found in roof collected rainwater including Salmonella, Campylobacter, Giardia, and Cryptosporidium. The likely sources of these pathogens were fecal material deposited by birds, frogs, rodents and insects, either in the gutters or in the water tank itself. However, detailed event studies undertaken in South East Queensland (SEQ) in Australia have shown that discarding the first 1(one)mm of runoff with a First flush device usually does little to substantially reduce the event mean concentration of bacteria and heavy metals entering a rain water tank during a storm event.

Simple illustration:
For designing a storage tank for rainwater harvesting system for a five-member family with typical Assam type house (sloped roof) having roof area of 100Sqm in Assam, following steps are adopted.
1) The Annual Rainwater harvesting potential would be:-
   =Area of Catchment (A) X Average rainfall(R) X Run-Off Co-efficient

= 100M² X 2.00mtr X 0.90
= 180 cum or 1, 80,000litres
(This is 3.50 times the harvesting potential for the same roof area of a house in Delhi).

2) Determination of tank capacity:-
This again depends on where the house is situated. i.e.If the RWH system would act as a stand alone source during dry period or act as a supporting system. Say, for designing a system to act independently as a single source for supplementing cooking & drinking needs of the family for dry season. Number dry days =180 day (October through March).

Quantity required for a five-member family = 180 X 5 X 10ltrs. = 9000litres.

The size of the tank would therefore be 3.00X3.00X1.20mtrs (with 0.20mtr Freeboard).

House hold rain water harvesting system near Aizawl, State of Mizoram, India, (Photo: Author)

Pronounced benefits of rain water harvesting systems where mark increase in ground water levels has been noticed:-

Sl. No.	Location of RWH systems.	Month/Year	Pre-monsoon depth of Ground Water Table (GWT)(Mtrs)	Post monsoon depth of Ground Water Table (GWT)(Mtrs)
1	Jamia Hamdard University, New Delhi	May'2002 May'2003	47.50	45.00
2	Janki Devi Mahavidyalaya, New Delhi	May'2002 May'2003	35.80	25.00
3	Panchshila Co-operative Housing Building Society Ltd, New Delhi.	May'2002 May'2003	28.50	27.60
4	The Sri Ram School, New Delhi.	May'2002 May'2003	40.00	38.10

Source: A Water Harvesting Manual for Urban Areas. "Centre for Science and Environment"

The benefits are not limited to an increase in the available quantity of water for usage. It also helps to improve the water quality.

Management by recharge of Ground Water:-
1. The ground water recharge helps to increase the yield of well & hand pumps.
2. The recharge of ground water dilutes the salts and harmful chemicals present in it thereby improving its quality considerably.
3. The surplus run-off from the catchment which otherwise would have drained-off is put to beneficial use.
4. There is effective rise in the water table in the recharged area; this in turn reduces the energy of lifting through pumps and thus less consumption of electricity and overall gain in economy.
5. As the water remains under ground, no loss due to evaporation is expected as in case of surface storage.
6. Large storage structures are avoided thereby cutting in costs of introducing such storage reservoirs.

Following situations warrants artificial recharge of ground water-
- When the ground water in the neighborhoods declines due to over exploitation.

- When the availability of ground water from the existing well diminishes during the lean period.
- When the ground water quality is poor with no alternative sources of water in the vicinity.

Following methods are employed to recharge the ground water at individual and community level:

Ditches & Furrows – The water from a natural stream is diverted through channels or ditches in the upstream of the natural stream. These ditches or channels are connected to furrows or gullies joining them at right angles. The water from the stream then spreads to theses furrows through the ditches covering a large area and in process recharges a wider catchment. The surplus water from the channels is passed back to the stream at the downstream end where the ditches or channels are again connected at some lower level.

Channel Spreading: – When a stream or a rivulet passes through a wide valley with lesser discharge and gentle slope. L-shaped bends of earth or impermeable mud is introduced within the stream width to increase its effective length so that the water passes through a longer path length in certain period of time. This helps to increase the detention time of the stream and in process, recharge the stream bed.

Percolation tank: – These are the artificially created surface water bodies to facilitate the percolation of the impounded surface run-off to recharge an aquifer beneath the tank. These are constructed in areas of hard rock with limited water holding and water yielding capabilities. A check dam is constructed across a small stream to create a pond filled with impounded surface run-off from upstream catchment. These are widely used in states of Maharashtra, Andhra Pradesh, Madhya Pradesh, Tamil Nadu and Gujarat in India. Some of the constructional features of percolation tanks are-

- Percolation Tanks are normally constructed in a terrain with fractured and weathered rock for speedy recharge of underlying aquifers.
- The aquifers to be recharged have sufficient thickness of the permeable vadose zone so as to accommodate the recharge. This vadose zone should normally be at 3(three) metres below the ground level to minimize the possibility of water logging.

- The area submerged due to the impounding of water should be an uncultivated land.
- The aquifer zone should extend up to the beneficiaries and wells or hand pumps should be provided in the area benefitted. The density of such well should be 3-5 per SqKm.
- The catchment should have light sandy soil so that silting of tank bed is avoided.
- The yield from a low catchment area is generally 0.44 to 0.55 MCM/SqKm. Accordingly, the catchment area for small tanks varies from 2.5 SqKm to 4.00 SqKm and for larger tank, it is 5 to 8SqKm.
- Depending upon the percolation capacity of the underneath formation of a tank, the percolation tank is designed for a storage capacity of 2.25MCM to 5.65MCM. The design capacity should normally be limited 50 percent of the total utilizable run off quantity.
- The height impounded water should be between 3mtrs to 4. 5mtrs.It is desirable that the storage is exhausted by month of February since evaporation losses increases from February onwards.
- The check dams to impound water in a percolation tank is generally made up of a mixture of soil, silt, loam, clay, sand and gravel, which are mixed and laid layers by compacting each layer before laying another on top to achieve sufficient water tightness. The dam is provided with free board and sufficient length of waste weir to avoid any over topping in the dam section.
- A properly located, designed and constructed percolation tank can reach efficiency ranging 78 percent to 91 percent with respect to ground water recharge. Generally, the zone of influence on the downstream side extends up to 1Km.

Recharge shafts:- Ground water can also be recharged using recharge shafts. The recharge shafts are one of the most effective structures to recharge an aquifer directly. The water can be directly fed into these recharge shaft to recharge and lift the ground water table. Following guide lines can be adopted in construction of recharge shafts.
- If the ground stratum is non-caving in nature, the shafts can be dug manually.

- If the available stratum of soil is caving in nature, a layer of boulders can be provided to support the surrounding soil and prevent from possible collapse.
- If the water from the source contains silt, the shaft is provided with layers of sand, gravel and boulder. The sandy layer at top has to be cleaned periodically to prevent clogging.
- The diameter of the shaft should at least be 2.00mtrs so that it can accommodate more quantity of water and also formation of eddies are avoided.
- The pipes through which water are fed in to recharge shaft, are to be lowered below the water level in the shaft, otherwise the air bubbles present, if any, can choke the aquifer.

There are two kinds of recharge shafts; (i) Vertical recharge shafts (ii) Lateral recharge shafts.

Vertical recharge shafts: These shafts are suitable for deep water level (up to 15mtrs from the ground level). These recharge shafts can be used when there is a layer of clay within 15mtrs depth. Depth and diameter of such shafts depend upon depth of aquifer and the volume water to be recharged. For diameter of 2-3 meters, the rate of recharge with inverted filters is usually 7-14 lps. Rate of recharge also depends on the silt content of water used for recharge and the aquifer material.

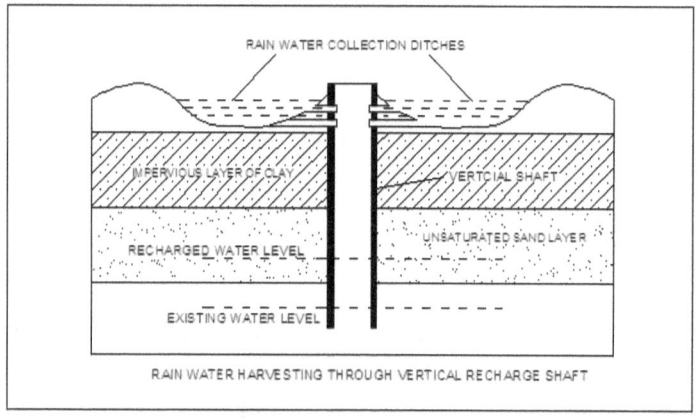

(ACAD representation: Author)

Sometimes, when the water level is deep and more than 15metres, vertical recharge shafts with injection wells are constructed. One or more injection well having diameters of 100mm to 150mm is constructed at the bottom of the recharge shaft. When the aquifer is overlain with an impermeable layer of clay, these injection wells are pierced through the impermeable layer to reach the potential aquifers to be recharged which may be at a depth of 3-15 meters below the water level in the shaft. The injection wells are filled with gravel to provide hydraulic continuity so that water is directly recharged in the aquifer. Depending upon the volume of water, the number of injection wells can be increased and recharge rate thereby can be increased. The efficiency of such vertical recharge shafts with injection wells are very high and can reach up to 15lps depending on volume of water and aquifer material being recharged.

Lateral recharge shafts: The lateral recharge shafts are constructed digging a trench of 2 to 3 meters depth and 2-3 meters wide, the length of which are decided based on the volume of water to be recharged. These recharge shafts are suitable when the permeable strata are available within 3 meters below the ground level and it extends up to the water level under unconfined conditions.

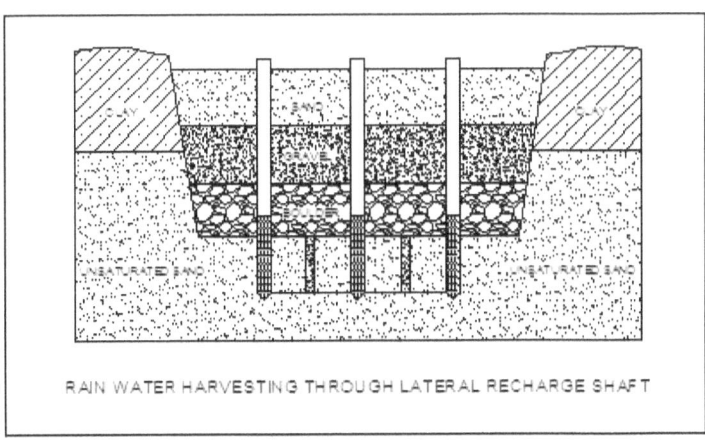

(ACAD representation: Author)

Advantages of recharge shafts:
1. The recharge of ground water through such shafts is fast and gives immediate benefit to the users.
2. It is simple to construct and maintain.

3. The loss of water through evaporation is avoided.
4. It needs less area for construction, so land acquisition cost is minimal. Disused household dug wells can be easily converted to recharge structures with little investments.

Gabion Structures: This is a kind of Check dam laid across a stream to retain water. This Check dam structure is made of locally available boulders and is netted with steel wire mesh. The height of the dam is kept at about 0.5 meters and generally they are laid for width of stream limited to 10-15 meters. The excess water overflows the gabion structure and water retained helps to recharge the ground. These are generally constructed in Maharashtra, Madhya Pradesh, and Andhra Pradesh states of India. The silt on the upstream side of the dam enters the interstices of the boulders and seals the dam making it further impermeable.

TRADITIONAL RAIN WATER HARVESTING SYSTEMS

The archaeological studies show that India's Great Rann of Kutch has several reservoirs to collect rain water runoff in the extremely dry regions of Dholavira, dating back to 300 BC. Kautilya's Arthshashtra, which was written in 300BC, has a mention of irrigation by rain harvested water through community participation. The Junargarh Inscription dating back to 2nd century AD informs about the restoration of Sudarsana Lake, which is not seen now since 9th century AD. The Vijayanagar tradition of Vijayanagar Kingdom (1336-1564 AD) laid much emphasis on the development of irrigation and water harvesting for improvement of agriculture. The king Krishna Deo Rai (1509-1530AD) etc emphasized that prosperity of the empire depended on the construction of irrigation channels and water tanks. The Gond tradition emphasised on the repairs of channels, embankments, distributaries, lakes or talabs etc. The Rani Talab of Jabalpur is a glaring example of water harvesting tradition of Gond tribes. Raja Bhoj of Bhopal built the largest artificial lake covering 65,000 acres in India. The lake was fed by streams and springs. The melting glaciers and snows are the only sources of water for the people residing in Ladakh region. The people of this region make intelligent use of their limited resources and make agriculture possible in this dry and barren land. The snow and ice melt slowly through the day and water is available in the streams only in the evening, when it is too late for irrigation. The water in the streams is hence led by channels to storage tanks and used the next day. These storage tanks are called "Zing". Apatani tribes in Arunachal Pradesh practice another system of water conservation, where the stream water is blocked by constructing a wall of 2 to 4mtrs high and 1mtr thick near forested hill slopes. This water is taken to agricultural fields through channels. The valleys are terraced in to plots separated by 0.60mtrs high earthen dams with inlet and outlets to the next plots that help to flood or drain the plots as and when required. Another tradition of water conservation is "Zabo", the term Zabo means impounding run-off. Zabo tradition is practiced in Nagaland. When rain water falls on terraced hill slopes, the runoff is collected in ponds in the middle terrace. The runoff then passes through slopes and finally reaches the paddy fields, at the foot of the hills. This serves not only in irrigation of the paddy field, but the fertility of the crop field is also improved. A very popular tradition of water harvesting in the north eastern region of India is the Bamboo drip irrigation, where rapidly flowing water from the streams and springs is captured by bamboo pipes and transported over hundreds of metres to drip irrigate black pepper cultivations in Meghalaya state of India. Ahar-Pyne is a traditional flood water harvesting system indigenous to South Bihar and in Palamu of Jharkhand. The Ahar is the catchment basin embanked on three sides, while the fourth side is a natural slope. Pynes or artificial channels start out from the river and meander through fields to end up in an Ahar. Johads are another popular traditional water harvesting technique where small earthen check dams that capture and conserve rainwater, improving percolation and ground water recharge. So far, some 3000 Johads have been constructed across more than 650 villages in Alwar district of Rajasthan. This has resulted in a general rise of ground water level by almost 6 metres and a 33 percent increase in the forest cover in the area. Five rivers that used to go dry immediately following the monsoon have now become perennial. One such river is Arvari, which has come alive. Yet another popular age-old Indian tradition of water harvesting is the construction of Khadins. A Khadin also called Dhora is based on the principle of harvesting rainwater on farmland and subsequent use of this water saturated land for crop production. Khadins were first designed by Paliwal brahmins of Jaisalmer, western Rajasthan in the 15th century, this system has great similarity with the irrigation methods of the people of Ur (present Iraq) around 4500BC and later of the Nabateans in the Middle East. A similar system is also reported to have been practiced 4000 years ago in Nagav desert and in South Western Colorado some 500 years ago.

Tanka: Tankas or Kunds are the traditional rain water harvesting structures widely constructed in some desert villages of Bikaner and Jaisalmer districts of Rajasthan in India. The Tanka system of conserving water for future use has also been in existence since many centuries in the pilgrim town of Dwarka. These are the traditional domed shaped underground storage tanks to store rain water for future use in drinking or domestic purposes. Tankas were often beautifully decorated with tiles which also helped the water to keep it cool. These structures are constructed for individual households as well as for village communities. For individual households, the rain water from the roof or terrace is directed towards an opening in the floor which surrounds the Tanka with a slope of 3% towards its inlet. These Tankas or the Kunds are generally circular in plan. They are constructed in stone masonry with cement - sand ratio of 1:3. For individual Tankas, the capacity of the Tankas are limited to 21Cum-59Cum, with diameter of such structures being 3mtrs to 4.22mtrs while the depth is kept equal to the diameter of the Tanka. For Tankas constructed for village communities, the diameter of the structure is kept at 6.00mtrs and the height, same as its diameter. The capacity of such Tankas is generally limited to 200Cum. The catchments around the Tankas or Kunds are given a slope of 3% (3cm in 1 metre) towards the inlet of the Tanka. The catchments are treated with gravel, pond silt, lime concrete or soil cement mix etc. to increase the rain water collection in the Tankas. As the cost of treating a large catchment surrounding a Tanka becomes uneconomical when the number of beneficiaries or the users are less, the treated surface is limited to a circular strip of 12mtrs surrounding such structures. This provides the requisite amount of water to fill the Tanka or the Kund. The remaining water is collected from the natural catchment surrounding the treated area. The Tankas with capacity of 21Cum built for individual households should be built in an open area 10mtrs x 10mtrs. The area where the Tankas or the Kunds are built should be free from human activities or cattle grazing during the monsoon so as to prevent any pollution of rain water collected in the Tankas. The Tankas built for village communities should be constructed in an open area of 30mtrs x 30mtrs.

Planning and Designing of Tanka

Water requirement: A Tanka built for an individual household and having a capacity of 21Cum is generally adequate to meet the drinking water requirements of a family of six persons for one year. The Tankas built for village communities, partly meets the water requirement of village communities but the main purpose of such structures is to conserve the available water and distribute it efficiently.

Water availability: In case of untreated catchment around the Tankas, some part of rain water is lost due to evaporation and seepage in to the ground. The loss varies with the intensity of rainfall. The losses are high when the rainfall is low and vice versa.

Yield from 1 hectare of natural (untreated) catchment.

Total monsoon rainfall in mm	Good catchment		Average catchment		Bad catchment	
	% of utilizable rainwater	Utilizable rain water (Cum)	% of utilizable rainwater	Utilizable rain water (Cum)	% of utilizable rainwater	Utilizable rain water (Cum)
20	0.080	0.16	0.06	0.12	0.04	0.08
40	0.130	0.52	0.095	0.39	0.065	0.26
60	0.245	1.47	0.173	1.10	0.122	0.73
80	0.410	3.28	0.307	2.56	0.205	1.64
100	0.700	7.00	0.525	5.25	0.35	3.50
120	0.900	10.80	0.675	8.10	0.450	5.40
140	1.122	17.15	0.918	12.86	0.612	8.57
160	1.625	26.00	1.218	19.50	0.812	13.00
180	2.120	31.86	1.590	28.62	1.060	19.08
200	2.700	54.00	2.025	40.50	1.35	27.00
220	3.260	71.72	2.445	53.79	1.63	35.86

Good Catchment:- Hills or plains with little cultivation & moderately absorbent soil.

Average Catchment:- Flat partly cultivated stiff gravely / sandy absorbent soil.

Bad Catchment:- Flat and cultivated sandy soils.

Run off from treated catchments (30m diameter) for rainfall range 130mm to 317mm:

Sl No.	Treatment	Percentage of utilizable rain	Utilizable rain (Cum)
1	Bantonite 20% mixed	51-87	46.80-194.84
2	Cement 8% mixed with soil 1.25 cm thick	23-41	21.12-91.82
3	Mud Plaster (Local) 1.25 cm thick	38-67	34.90-150.00
4	Lime concrete 5cm thick	48-74	44.00-165.73
5	Sodium Carbonate spary @ 1Kg/10Sqm over 1.25 cm thick compacted tank silt.	63-92	57.86-206.00
6	Mud Plaster with mixer of mud, Wheat husk (Bhusa) and Jantha Emulsion (a kind of ashphalt) (95:3:2)	49-79	45.00-176.92
7	Well dressed and compacted without treatment	30-57	27.55-127.85

Constructional features

For 21Cum capacity (For domestic use):
- Excavation for foundation: Diameter of excavation -3.90mtrs and Depth of excavation-3.9mtrs.
- Foundation Slab: Slab of 3.90mtrs Diameter and 250mm thick, constructed in Cement Concrete (C.C) 1:3:6.
- Vertical wall: Constructed in Cement Concrete (C.C) 1:2:4 and of 250mm thickness with 5mm thick cement plaster.
- Tanka Cover: Stone slab roof at a height of 1.00mtr above ground level.
- Apron: Apron around the Tanka is 1.00mtr wide and of 100mm thickness built in Cement Concrete 1:3:6.
- Catch Pit: A deep catch pit of size 1mtr x 0.25mtrs at the bottom of the tank is provided to ensure proper cleaning of the Tanka.
- Inlets & Outlet: 3 Inlet and 1 Outlet of size 0.60mtrs x 0.30mtrs is provided at the apron level. Inlets and Outlets are covered with iron grates.

- Top Opening: An opening at top of size 1mtr x 1mtr is provided for drawing water.

For 200Cum capacity (For community use):
- Excavation for foundation: Diameter of excavation -9.150mtrs and Depth of excavation-6.32mtrs.
- Foundation Slab: Slab of 9.15mtrs Diameter and 250mm thick, constructed in Cement Concrete (C.C) 1:3:6.
- Vertical wall: Constructed in Random Rubble masonry in 1:3 cement-sand mortar and having 1.45mtrs thickness at bottom and thickness gradually reducing to 0.38mtrs at top.
- Tanka Cover: Stone slab roof over R.S joists of size 0.20mtrs x 0.15mtrs at a height of 1.00mtr above ground level.
- Apron: Apron around the Tanka is 1.60mtrs wide and of 115mm thickness built in Cement Concrete 1:3:6.
- Catch Pit: A deep catch pit of size 0.915mtr x 0.23mtrs at the bottom of the tank is provided to ensure proper cleaning of the Tanka.
- Inlets & Outlet: 3 Inlet and 1 Outlet of size 0.60mtrs x 0.30mtrs is provided at the apron level. Inlets and Outlets are covered with iron grates.
- Top Opening: An opening at top of size 0.90mtr x 0.9mtr is provided for drawing water.

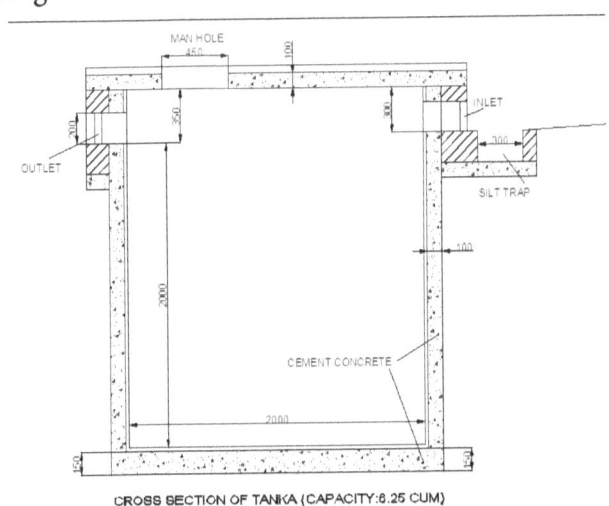

CROSS SECTION OF TANKA (CAPACITY:8.25 CUM)

(ACAD representation: Author)

Advantages of Tankas:
i. Since the rain water flows in to inlet of the tank through a gently sloping treated catchment, comparatively clean water is collected in the Tanka.
ii. The Tanka being covered with a roof minimizes the evaporation losses.
iii. This method of rain water harvesting is comparatively cheap and environment friendly.

Limitations:
i. Special attention is needed for protection of catchment from cattle grazing and pollution by human activities.
ii. Chances of reptiles, insects and even small animals entering the Tanka through the inlets are not ruled out.
iii. They are not useful in areas of steep slopes and clayey soil surface.
iv. Tankas need periodical cleaning.

Khadin: This is a very old system of rain water harvesting widely followed by the Paliwal Brahmins of Jaisalmer of Western Rajasthan in India in the 15th century. The similar system has been in use in Ur (present Iraq) some 4500 years BC and also practiced in Negav Desert in the Middle East region around 4000 year ago. A Khadin area or the cropped area is selected for construction of the Khadin bund or a low earthen embankment which retain the surface run off from the upland catchment within the Khadin or the cropped area. The Khadin area so selected is in proximity to a high catchment area with adequate run off to fill the Khadin area during the rainy season. The ratio of the Khadin area to such high or upland catchment area generally varies from 1:2 to 1:15. The length of the Khadin bund is decided on basis of the boundary of the cropped area and is generally 100mtrs to 300mtrs long. The top soil where the bund is constructed is first excavated to a depth of 15cms below the ground level. The excavation is then filled up with impervious clayey layer matching the foundation width of the bund or the earthen dam. A profile of the bund is erected at every 20mtrs of the earthen dam. The earthen embankment is then constructed by filling in layers of soil of 30cms thickness each and compacting with sheep foot roller or by hand ramming. Khadin area just upstream of the earthen embankment or the dam is leveled to increase the water spread of the impounded run off. An overflow spillway is constructed with cement concrete to drain way the excess water to the downstream side. The downstream side of the spillway is covered with boulders to prevent erosion. Sometime a scour pipe is provided through the embankment to drain away the stagnant water on the upstream side. After the construction of the earthen embankment, it is planted with grass to give proper stability and avoid erosion. One or two dug wells are also excavated on the downstream side of the embankment for consumptive use of the recharged ground water.

Nadi: Nadis are the excavated village ponds with side embankments used for conservation of water collected from the adjoining catchments for drinking purposes in some desert villages in India. The first recorded masonry Nadi was constructed in 1520 AD during the regime of Jodhaji near Jodhpur. To minimize the cost of construction, low lying areas where excavation requirement is minimum, and where rain water would get collected naturally from the adjoining catchments are selected for construction of a Nadi. The Nadis or small ponds which cater to a small village would supply drinking water needs to the village for 2 months to a year after the rainy season depending upon the catchment characteristics, intensity and duration of the rain fall. These ponds are now given HDPE lining by using HDPE sheets at the sides of the embankment to prevent loss of water at the bottom or the sides. The inlet is provided with silt trap, so that silt is prevented from entering the pond through the inlet. Barbed wire fencing is provided around the Nadi to prevent pollution due to any ingress of floating material. The inlet is stone-pitched.

Ahar-Pyne: The average annual rainfall in Bihar varies from 1000mm in Patna to 1600mm in eastern parts of the states. The northern districts of Saran and Southern part of Muzaffarpur lying in the north of the Ganges, even though considered dry, are to some extent protected by water-retaining capacity of the region's new alluvial type of soil. The southern part of this dry zone, comprising districts of Patna, Gaya, Shahabad, south of Monghyr and south of Bhagalpur, commonly known as south Bihar is composed of mostly old alluvial soil having very little water holding capacity and thus dries up quickly as soon as the rain stops. This area having marked slope from south to north induces quick flow of rain water runoff from the catchments thus making it difficult for adjoining soil strata to retain substantial portion of the runoff water. The natural conditions in south Bihar is not favourable to regular cultivation. It is the ingenuity of the locals in this part of the country, whose innovation of the age-old rain water harvesting structures like Ahars & Pynes, not only made them survive but also made the region one of the most populous tracts in the world over two millenniums.

South Bihar is bounded by Chhotanagpur plateau in the south and the Gangetic valley to the north. This region has a gradient of approximately one metre per Kilometre from south to north. An embankment having height of one or two metres is built on the lower side of a terrain usually on north to impound water that would flow naturally during and after a rainfall from the south. From the two ends of this embankment, two other embankments are constructed which projects towards the south and gradually reducing its depth meeting the natural ground on the higher southern slope. This constructed array of embankments closing on three sides which impounds water within and receives rain water runoff from the open fourth side, i.e. catchments of the higher southern side, is called an Ahar. Unlike tanks, beds of ahars are not dug out.

The rivers meandering across the hills of Chhotanagpur plateau in south Bihar are mostly having sandy beds. Most of these small rivers remain dry for most of the year but once the rainy season starts, these rivers turn into swollen torrents and flow rapidly from south to northern Gangetic valley. As soon as the rains stop, these swollen rivers again become dry with exposed beds of sand as most of the water is carried off north or percolates in to dry beds of sand. Small artificial streams are dug out from the points facing the current of these rivers and are led to nearby or distant agricultural field. These artificial streams are called Pynes. Some of these pynes are 20-30 kilometres long and irrigates agricultural lands adjoining dozens of villages in south Bihar. As the sandy bed of the source river is at a higher elevation than the water level of these pynes after it is carried to certain distances owing to natural south to north gradient of the terrain in these parts of the country, the pynes need little or no excavation to feed water from these artificial streams to the adjoining irrigated agricultural fields. However, sometimes some artificial embankments are constructed across these pynes to raise and divert water from these pynes to the adjoining agricultural lands. These pynes are also led to ahars to make available the required volume of water in the ahar from the surplus of these pynes.

The system of irrigation with ahars and pynes were highly developed in the Gaya district of south Bihar. The first Irrigation Commission (1901-1903) had noted that more than half the total area of the district, comprising of some 1,670,000 acres were irrigated through this age-old system of irrigation. In fact, ahars and pynes constituted nearly three-fourth of all forms of irrigation in south Bihar.

Ahar-Pyne irrigation in Mukhdumpur, Jehanabad, Bihar, India (Photo: Author)

Bamboo Drip Irrigation: This is one of the most ingenious techniques of water usage in plantations in hill slopes and valley lands of Meghalaya state in India. A network of bamboo cut to suitable lengths of 8-10 feet and in places even longer is used to carry water from the numerous natural streams existing in this state. The bamboos are slit longitudinally in to two halves and knots removed. These are then arranged in stepped criss-cross series to carry water from suitable points on the streams to distant plantations and in process saving enough water that otherwise have gone waste. Throughout its journey, the clean water travels under gravity in its natural surroundings.

Bench Terrace Cultivation: This is mostly practiced in the state of Meghalaya in India and also widely prevalent in other north east Himalayan regions. The rain water emerging from forests in the upper reaches of the hills are channelized to pass through a series of terraces, sized and leveled for rice plantation. The water is allowed to flow from top to bottom terraces in stepped fashion, filling each terrace with water depth of 5-8 centimeters. The edges of the stepped terraces where rice cultivation is done is embanked with earth to retain water leaving a small outlet in the embankment to pass water to the next lower terrace after ensuring desired water depth of 5-8 centimeters in the preceding upper terrace from where water emerges. This type of conservation measure is predominantly used for rice cultivation. However, in rain fed conditions, sequencing of cultivation is done with upper reaches of the slopes being used for maize, bean, and potato cultivation and then water passing on to lower terraces with rice cultivation which require more water.

Bench Tearrace Cultivation near Umiam, Meghalaya, India (Photo: Author)

Erys: Erys are age old irrigation and water harvesting system of South India more prominently in Tamil Nadu. An Ery is a water reservoir bounded on three sides by earthen bunds or embankments and the fourth side is open to catchments from where water is collected in the Ery through gravity. As a normal practice, middle of the bund is the deepest portion and water depth decreases as one moves away from the middle to side flanks of the bund. Apart from water harvesting, one main function of Erys is to irrigate cultivation fields. The Erys are designed to irrigate certain areas of cultivation downstream by supplying water through channels, these areas under irrigation through water supplied from Erys are called ayacuts of the Ery. The water to these ayacuts passes through valves or sluices located in the bunds of the Erys which are controlled from top of the bunds.

One or more overflow weirs are constructed on one of the flanks of the bund, so that when water reaches certain height within the Ery, it overflows and reaches another Ery in some other village downstream through gravity. This way number of Erys is interconnected with overflow water from one upstream Ery reaching another downstream and does collectively irrigating a large area under cultivation. Ery irrigation accounts for more than 4 million hectors (mha) of cultivated land in India.

Major uses of Erys

Erys in irrigation of paddy cultivation: These reservoirs serve as a supplementary source of irrigation for rain dependent crops such as paddy. The first season crop is called samba, which begins in Tamil month of Adi (July-August). The samba varieties of crop are usually of six months duration. Towards the end of this crop, Erys usually contain enough water to bring this crop to maturity. If rains are sufficient to fill these Erys, there is usually a second crop of paddy.

Erys as flood control and water storage device during heavy rainfall: During the north east monsoon in Tamil Nadu. Sudden heavy rainfall, in absence of Erys, would cause huge quantity of water flow down the Eastern Ghats carrying along with it all the silts in the Bay of Bengal Sea. The chain of Erys helped prevent sudden flash floods by passing overflow waters from upstream Erys to fill the downstream Erys and in process filling a large number of Erys and storing rain water and lowering the speed of flowing water that would otherwise cause breaches and siltation along its path. Erys, thus prevent flooding of a large area and also store rain water which would have otherwise gone waste and flow into sea. It is imperative that the chain of Erys is maintained properly by de-silting to keep the storage level at maximum. If not properly maintained, siltation of Erys would lower its reservoir capacity, thus making bunds of such Erys vulnerable to breaching and failure during heavy rainfall, resulting in a catastrophic effect on the Erys downstream leading to their successive failure and flooding of a large area.

Erys for recharge of ground water: A well maintained Ery would store water for a long duration and ensure water availability during lean period of low rainfall. It is recorded in the north Arcot District manual that towards the end of 19th century, the Duri-Mamandur Ery in north Arcot district of Tamil Nadu stored water for 15 months. This is noteworthy that during that period Erys were already on the decline due to poor maintenance. Thus, a properly maintained Ery would ensure storage of water during the period of scanty rainfall and the standing water also helps to recharge the ground water and help maintain water level at individual and community wells in villages at the same time

TRADITIONAL WATER HARVESTING SYSTEMS AROUND THE WORLD

Asia: In the middle east, archaeological evidence of water harvesting structures appear in Jordan, Israel, Palestine, Syria, Iraq, Negar and Arabian Peninsula region (mainly Yemen), the oldest being believed to have been constructed over 9000 years ago. In Baluchistan, Pakistan, two water harvesting techniques have a long tradition: the "Khuskaba" macro catchment system and "Sailaba" system, which utilises floods.

Africa: In North Africa, water harvesting has a long tradition too, and is still used extensively in Morocco, Tunisia and to a lesser extent in Algeria. Traditional techniques of water harvesting have also been reported from many regions of Sub-Saharan Africa. Like "Lag" and the Gawa" systems in Somalia. Various types of other harvesting systems like "Hafirs" in Sudan and "Zay" system in West Africa were also popular.

Flood Water Harvesting

It may also be termed as "Large catchment area harvesting" or "Spate Irrigation". There are two forms of flood water harvesting,
i). Harvesting within the stream bed where the water flow is dammed, as a result, inundates the valley bottom of the flood plain. The water is forced to infiltrate and the wetted area can be used for agriculture or pasture improvement.

ii). Harvesting through flood water diversion where the wade water is forced to leave its natural course and conveyed to nearby cropping area.

Flood Water Harvesting in select North African and Middle East Countries	
Region	Area under Flood water Harvesting (x 1000 hectares)
Algeria	110
Eritrea	16
Morocco	165
Pakistan	1402
Somalia	150
Sudan	46
Tunisia	30
Yemen	98

(Source: FAO 1997)

Quanat Systems: A Quanat is a horizontal tunnel that taps underground water in an alluvial fan without pumps or equipment, brings it to surface so that the water can be used. Quanat tunnels have an inclination of 1-2percent and length of upto 30 Kms. The origin of Quanat technique is in Persia, where it was developed some 3000 years ago. The knowledge spread to the neighbouring countries and later to whole of North Africa and even to Spain. Many Quanats were constructed in India too and are still in use under different names in Kerala (Surangam) and Madhya Pradesh (Bhandare System). Though new Quanats are seldom built today, many old ones are still maintained and deliver steadily water to fields and villages. In Morocco, many of the large Oases of the south receive a considerable amount of their irrigation water from Quanats; rest is harvested flood water.

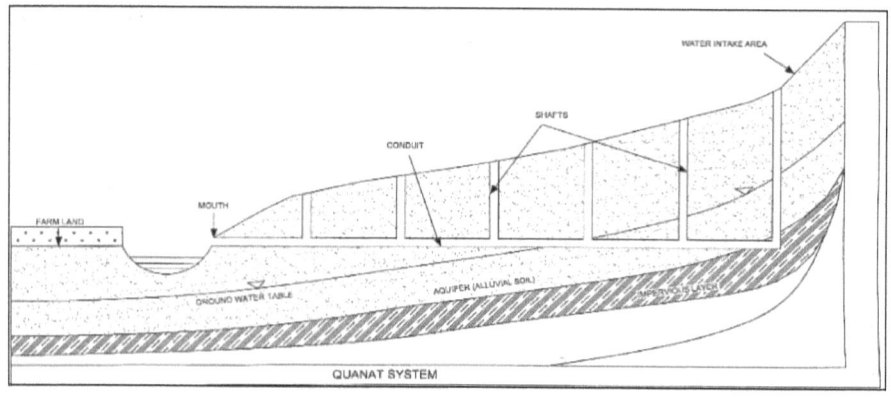

Quanat System of conveying water by gravity to the ground
(Natural Academy of Science 1974), (ACAD representation: Author)

Sand Dams: Sand dams are usually built in the upper and middle courses of seasonal sandy river valleys (also known as wadis). They are constructed in transition zone between hills and plains where gradient of the river bed is small between 0.2 percent to 4percent, but in extreme cases depending upon site conditions they have also been built on slopes ranging 10 to 16 percent. These dams are built on bed rock or highly compacted sub-soil.

Constructional features: They are constructed on rocky bed with foundations going below till it rests on rocky surface or the stable impervious layer. These dams are reinforced concrete structures constructed across a sandy river. The sand accumulates slowly in stages after every seasonal rain behind upstream face of the dam and rain water passes on to the downstream side till the dam fills with sand having 25 to 40 percent of the volume with water.

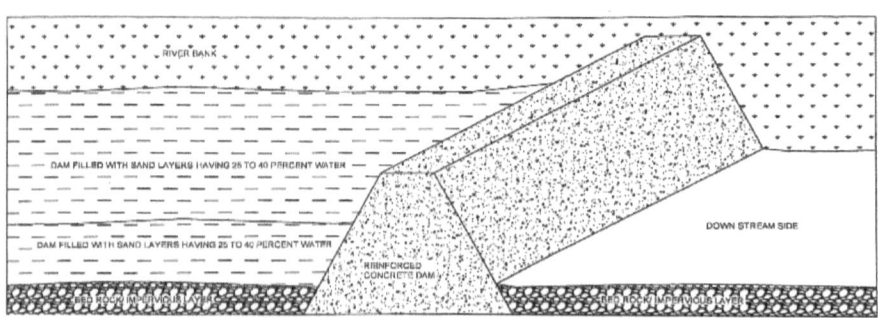

Sand Dam, (ACAD representation: Author)

Benefits of Sand Dams:
1. These are known to be lowest cost form of rain water harvesting system costing 3 to 100 times lower than other technologies. It lasts and gives benefits for more than 50 years with very little maintenance and operational costs.
2. It is mostly owned and managed by local communities which are keys to its successful operation.
3. They help in saving time of the members of the communities which otherwise walk miles to fetch water for their daily needs which are often unsafe, during drought in semi-arid regions.
4. They provide a clean, secure and local year-round water supply in water scarce environments. The sand on the upstream side effectively acts as a slow sand filter and supply potable water for the local communities.
5. As the water is held under the sand, evaporation losses are curtailed to maximum.
6. Areas, where the sand dams are in use, benefits in terms of health, education in terms of school attendance and significant increase in house hold income and food security were noticed.

More than 500 sand dams were set up in Kitui region in eastern Kenya which helped in receiving benefits in terms of water storage, domestic use, and for feeding and sustaining livestock.

Reducing water demand through Nitrogen fixing trees

Nitrogen fixation is a process in which nitrogen (N_2) in the atmosphere is converted in to ammonia (NH_3). The atmospheric nitrogen is present in air in stable diatomic form N_2. Almost 80 percent of air is made of nitrogen, this converts to some 6400Kg of nitrogen per hectare of land. The nitrogen fixing trees (NFT) are deep rooted and thus have access to nutrients in the sub soil layers. The extensive root system of nitrogen fixing trees helps in stabilizing the soil. The roots create channels of aeration within the sub soil pores. The leaves of the nitrogen fixing trees, dropping on the ground, fertilize or nourish the soil further. These are called the pioneer plants because they survive in harsh conditions and in soil having low fertility. These plants or trees add and accumulate nutrients in soil on which they grow and make the soil fertile and favourable for other non-pioneer plants to grow. The atmospheric nitrogen is converted to ammonia by a group of bacteria called Rhizobia. Hundreds of colonies of such bacteria are found in root nodules of nitrogen fixing trees. The roots provide energy to these bacteria to work on the atmospheric nitrogen and these bacteria in turn provide the roots with necessary nitrogen for growth. The biological nitrogen fixation was discovered by Germen agronomist Hermann Hellriegel and Dutch microbiologist Martinus Beijernick.

The reaction for biological nitrogen fixation is

$$N_2 + 8H^+ + 8e^- \rightarrow 2NH_3 + H_2$$

Agro forestry is a blend of agriculture and forests where crops cover 80 percent or more areas and forest cover 20 percent or fewer areas. Basically, agro forestry is a land management system rather than water management. However, agro forestry can enhance crop production through nitrogen fixing trees planting, with water utilization remaining more or less the same, thus enhancing food security. Nitrogen fixing plants like Faidherbia sheds its nitrogen rich leaves during rainy days when crops are grown, so they do not compete with the associated crop for nutrients, water or light. The leaves grow again during the dry season and provide shed for associated crops. In general, agro forestry with nitrogen fixing trees can help in following ways:

- It would help to reduce deforestation by providing fuel wood to farmers made available by nitrogen fixing trees.
- It enhances production of associated crops thereby increases food security. Certain Nitrogen fixing trees like TephrosiaVogelli and Gliricidiasepium planted with maize in Malawi region of Africa enhanced the production of associated maize crops to an average of 3.7 tonnes per hectare in a ten-year experiment compared to 1 tonne per hectare without nitrogen fixing trees or mineral fertilizers. Thus, agro forestry with nitrogen fixing trees helps to obtain maximum output from the same piece of land.
- Agro forestry incorporates several plant species within given land area. It creates a more complex habitat in addition to increasing land fertility. It supports wider range of birds, insects and other animals.

The nitrogen fixing trees in agro forestry system take up and store carbon at a faster rate than crops. Therefore, such blending of nitrogen fixing trees in agro forestry help reduce climate change.
- Agro forestry with nitrogen fixing trees has the potential to create extra income in the village community where in farmers can avail wood from fast growing nitrogen fixing trees to sell in local market or to wood-based industries in the vicinity.

Nitrogen fixing trees are blended in following ways in agro forestry-
Mulch Banks: Patches of nitrogen fixing trees are planted within the crop area to provide sheds to crops and distribute nutrients evenly.

Contour Hedgerows: The nitrogen fixing trees are planted in rows across a downhill slope inside the crop area. This helps to retard any siltation due runoff during rainfall. The rows of nitrogen fixing trees lying across the path of flowing rain water along the slope tend to reduce its speed and quantity and thus prevent siltation and also help in avoiding flash flood in the foot hill plains. It is reported that in Western Himalayan region, the Hedge rows planting reduced run off by 40 percent and soil loss by 48 percent under 1000mm rainfall condition.

Reducing water demand through Waste Water Management

Out of the estimated 40673 Km^3 of annual renewable fresh water, the total fresh water withdrawal on global basis in 1987 was 3240 Km^3. Thus globally, approximately eight percent of the total renewable fresh water was withdrawn in 1987. The per capita withdrawal of fresh water during the same year was 660 m^3/person/year, i.e. about 1800litres per capita per day. Of this, eight percent, i.e. 145 litres accounted for domestic use, twenty three percent, i.e. 416 litres per capita per day was used for industrial use and approximately sixty nine percent. i.e. 1247 litres per capita per day was utilized for agriculture purpose. It is estimated that the global population used fifty four percent of world's accessible supplies of renewable fresh water in 2000 and it is projected that by 2025, three quarters of world's renewable fresh water supplies will be utilized on yearly basis. Exponential growth in world population with surge in industrial and agricultural activities during the last two decades of 20th century and 1st decade of the 21st century has forced many countries to opt for waste water utilization to reduce their water demand. Rise in economic growth in Japan during 1960's saw rapid strides in industrial activities. There was a concomitant rise in ground water abstraction with increase in industrial activities. As a result, there have been numerous instances where ground water depletion caused land subsidence. Policy planners in Japan felt an urgent need to reuse water especially after the water shortage that hit Fukuoka in 1978. Subsequently, proportion of recycled water in Japanese industries reached 76.8 percent in 1994 with Chemical and Steel industries attaining a recovery rate as much as 80-90 percent. By 1997, Japan saw 1475 on site individual building and block wise reclamation and reuse systems providing recycled water in commercial buildings and apartments for Toilet Flushing, Car washing, Landscaping etc. In addition, there were 163 publicly owned waste water treatment plants (POWTs) across Japan. The total amount of reused water was 480MLD (million litres per day), which corresponded to about 1.5 percent of the total treated waste water.

Waste water can be utilized for the same purpose for which it was used previously or for another purpose after it has been treated. Government of India launched the National Program on Energy Recovery from Urban Municipal and Industrial Wastes with following primary objectives:

(i) Creation of conductive conditions and environment with fiscal and financial incentives to help promote, develop, demonstrate and disseminate and utilization of waste for recovery of energy.
(ii) Improving waste management practices through the adoption of technologies for conversion of wastes in to energy and
(iii) Promoting the setting-up of projects, utilizing wastes from urban, municipal, and industrial sectors.

Some of the water intensive industries where treatment of waste water is necessary before disposal are:
- *Paper Industry*: The water requirement of paper industry in India range from 250 cubic metres to 450 cubic metres per ton of production, whereas the water requirement in paper industries in some developed countries range from 60 cubic metres to 120 cubic metres per ton of production. Study on waste water recycle in paper

industry revealed that waste water recycle in paper industries varies from 8 to 48 percent with an average range of 21.4 percent, which facilitates reduction of fresh water requirement to the extent of 7 to 44 percent.
- *Tanning Industry*: Approximately thirty to forty cubic metres of water is used for processing raw hide in to finished leather. Quality of waste water from tanneries depends upon type of tanning process and capacity of the tannery. Tanneries in proximity to one another can put up common effluent treatment plant to reduce cost of treatment.
- *Chlor-alkali industry*: Mercury pollution is associated with chlor-alkali industry. The quantity of mercury bearing effluent generated, range from 0.14 to 36 cubic metres per ton of caustic soda produced. Combination of pretreatment followed by ion exchange process in conjunction with sand filtration and activated carbon filtration presents one of the best options for treatment of mercury bearing waste water.

Waste water use in Agriculture: The principles driving the use of waste water in agriculture are:
(i) Increasing water scarcity and stress, and degradation of freshwater resources resulting from disposal of waste water in to natural water bodies without pretreatment.
(ii) Increase in population and related increase in water foot prints.
(iii) A growing recognition of nutrient value of waste water.
(iv) The Millenium Development Goals (MDGs), especially eliminating poverty and hunger, keeping in pace the sustainability of environment.

Following health protection measures are advocated to protect the consumers of waste water irrigated agriculture product, farm workers and their families and among communities residing in the vicinity of waste water irrigated fields
(a) Waste water treatment.
(b) Crop restriction.
(c) Waste water application techniques that minimize contamination (e.g. drip irrigation).

(d) Withholding periods to allow pathogen die-off after the last waste water application.
(e) Hygienic practices at food markets and during food preparation.
(f) Health and Hygiene promotion.
(g) Produce washing, disinfection and cooking.
(h) Immunization.
(i) Disease vector and intermediate host control.
(j) Reduced vector contact.
(k) Access to safe drinking water and sanitation facilities at farm.
(l) Restricted access to irrigated fields and hydraulic structures.

Summary of health risks associated with the use of waste water for irrigation:

Exposure Group	Health Risks.		
	Helminthes infections	Bacterial/virus infections	Protozoal infections.
Consumers	Significant risk of *Ascaris* infection for both adults and children with untreated wastewater.	Cholera, typhoid and shigellosisi outbreaks reported from use of untreated wastewater. Seropositive responses for Helicobacter pylori (untreated); increase in non-specific diarrhea when water quality exceeds 10^4 thermotolerant coliforms/100ml	Evidence of parasitic protozoa found on wastewater irrigated vegetable surfaces, but no direct evidence of disease transmission.
Farm workers and their families.	Significant risk of *Ascaris* infection for both adults and children with untreated wastewater. Risks remain, especially for children, when wastewater treated to <1 nematode egg per litre; increased risk of hookworm infection in workers.	Increased risk of diarrheal disease in young children with wastewater contact if water quality exceeds 10^4 thermotolerant coliforms/100ml; elevated risk of *Salmonella* infection in children exposed to untreated wastewater; elevated seroresponse to norovirous in adults exposed to partially treated wastewater.	Risk of *Giardiaintestinalis* infection was found insignificant for contact with both untreated and treated wastewater; increased risk of amoebiasis observed with contact with untreated watserwater.

Exposure Group	Health Risks.		
	Helminthes infections	Bacterial/virus infections	Protozoal infections.
Nearby communities.	*Ascaris* transmission not studied for sprinkler irrigation, but same as above for flood or furrow irrigation with heavy contact.	Sprinkler irrigation with poor water quality and high aerosol exposure associated with increased rates of infection; use of partially treated water (10^4–10^5 thermotolerant coliforms/100ml or less) in sprinkler irrigation not found to be associated with increased viral infection rates.	No data on transmission of protozoan infections during sprinkler irrigation with wastewater.

Wastewater use for recharge of ground water

Quality of land and its preparation for recharge with wastewater:

Land for recharge should be such that it is not fertile enough to support useful crops owing to insufficiency in soil cover. However, the soil should be able to infiltrate sewage without causing any ponding or water logging or overflowing. The land having permeable substratum is suitable. Close soils, deep clays are not suitable. On the other hand, very coarse sand and gravel are also not ideal as they facilitate fast movement of wastewater in to the ground, which again may pollute the ground water. When natural bed of sand is available this can be converted to a natural filter by removing the top soil. The plot selected for recharge should be on lee ward side of the township to prevent any odour nuisance. The plot should be located in a place where there are minimum chances of floods. The plot of land should not have an excessive slope which would give rise to uneven spreading. The excessive slope would also increase the cost of expenditure for leveling.

The land used for recharge should be leveled with occasional removal of some surface soils. The land should be divided in fair number of plots with enclosing embankments so as to prevent entry of rain water, which may cause flooding. The embankments used should have stable slopes. The inlet arrangements should be so made that it gives even spreading of sewage over the entire plot of land. The wastewater should not fill all plots for recharge at one time, but, some plots should be rested while others are being used. If a subsoil stratum is too porous, adequate soil cover is necessary; otherwise, quality of drainage water may be affected. For effective infiltration, the effective size of soil could be between 0.2mm to 1.00mm.

Rate of Application of Wastewater for recharge

The effluent or wastewater used for recharge should be applied intermittently. The effluent should be spread uniformly to a depth of 10-15cm for each application. The number of applications of effluent over the plot of land depends on quality of soil stratum. When the substratum is permeable up to a depth of 1.5m to 2.00m below ground surface, 2 to 3 applications per day is possible. Where sandy soil is available, this can be converted to a natural filter by scrapping the top soil, in such case, the rate of application varies from 900m^3/ha/day to 4500m^3/ha/day. The degree of such application also depends on the quality of effluent. While lower rate is applicable for raw sewage, higher rate of application is achievable for fully treated sewage. In case of higher rate of application, the ground water table has to be fairly low. In most Indian cities where only primary treatment is done, the rate of application of effluent for recharge could be 1100m^3/ha/day to 1400m^3/ha/day. Imhoff and Fair have suggested a rate of 3500m/ha/day for un-cropped area. The loading rates and types of soil for some locations in the U.S are as under-

Sl No.	Location	Hydraulic loading rate(m^3/ha/d)	Soil Type	Type of Effluent
1	Whittier Narrows (Los Angeles).	3500	Sand	Secondary
2	Flushing Meadows. (Phoenix)	3000	Sand	Secondary
3	Santee (San Diego).	2230	Gravel	Secondary
4	Lake George (New York)	1200	Sand	Secondary
5	Calumet	950	Sand	Untreated
6	Ft.Dvens (Massachusetts)	800	Sand and Gravel	Primary
7	Hemet (California.)	900	Sand	Secondary
8	Westby (Wisconsin)	300	Sandy Loam	Secondary

Methods of Recharge

Recharge through ditches: This method is used when transmissibility of soil strata is low. Parallel ditches are dug on the ground enclosed by embankments at sides. The ditches are provided with slight gradient for movement of wastewater. The ditches are usually 0.6m to 1.2m wide and the height of embankment is kept at 0.3m to 1.00m. The clear distance between two ditches is usually 2.00m. The bed of ditches is formed with a layer of 15-25cm of washed gravels or crushed stones.

Recharge through ponds: When transmissibility of the soil is high, recharge with wastewater can be achieved through a single pond having a large size. The size of the pond is decided based on volume of wastewater for recharge.

Recharge through spreading basins: When soil strata is sandy or of porous nature, the recharge of ground water can be done by constructing shallow basins of rectangular shape having length of few hundred meters and spreading wastewater to a depth of 20-30cms.

Recharge with sprinklers: This method uses sprinklers to spread wastewater over the recharge basin. Land surface should have adequate vegetation to preclude any run-off. Systems with loading rates $146 m^3$/ha/d or more can adopt this method.

Recharge through wells or shaft: This method of recharge is employed where there is a pervious layer of soil underneath an impervious surface layer. Wells of diameter 150-300cms are dug through the impervious stratum to pass the wastewater to the under layer of pervious soil.

Water reuse in Residential and Public Buildings

In 1984, the "Ministry of construction" in Japan implemented "Guidelines for use of Non-potable purpose". These guidelines suggest water quality criteria for reclaimed water for use in Toilet flushing, Landscape irrigation, and for Ornamental/Environmental purpose for artificial lakes, waterfalls, streams etc.

Water Quality Criteria for reclaimed water as per guidelines for use of non-potable purpose:

Parameters		Toilet flushing water	Landscape irrigation	Environmental water
Criteria	Total Coliform (CFU/ml)	≤ 1000	Not detected	Not detected
	Residual Chlorine(mg/l)	Trace amount	≥ 0.4	-
Guidelines	Appearance	Not unpleasant	Not unpleasant	Not unpleasant
	Turbidity, unit	-	-	≤ 10
	BOD, mg/l	-	-	≤ 10
	Odour	Not unpleasant	Not unpleasant	Not unpleasant
	pH, unit	5.8 – 8.6	5.8 – 8.6	5.8 – 8.6

The water reuse systems for buildings can be classified in to (i) Closed Loop Recycling System and (ii) Open Loop Recycling Systems.

(i). *Closed Loop Recycling System*: The wastewater from a building or group of buildings is reused in the same building or group of buildings after treatment.

(ii). *Open Loops Recycling System*: The water from a cluster of buildings is led to a wastewater treatment from where the treated wastewater is utilized for different purposes.

(i). Closed Loop Water Recycling System is again used in (a) Individual Building Water Reuse Systems and (b) Block wide Water Recycling System.

   (a). *Individual Building Water Reuse Systems*: In large office building and apartment complexes, the water for toilet flushing consumes up to 20-50 percent of the treated quantity. As such the grey water from wash basins, showers, cooling towers are led to a wastewater treatment plant located generally at the basement of the building. The treated wastewater is then reused through a separate distribution system for toilet flushing, car washing and gardening etc.

   (b). *Block-wide Water Recycling System*: The individual buildings are connected through a network of wastewater pipes, which lead grey water from these buildings to a centrally located wastewater treatment facility. The reclaimed water is then returned to these buildings through another set of urban distribution network for reuse largely for Toilet flushing.

(ii) *Open-Loop Water Reuse System*: The reclaimed water is reused in offsite buildings or industries located at distant places. Some of these waters are also led to nearby stream to augment the stream flow.

A case study in Bangalore, India:
A wastewater recycle plant was installed at the rear side of Bangalore city (India) railway station to recycle and reuse wastewater for washing bogies. On an average, 70-80 trains per day are washed at city railway station at Bangalore. The total water requirement of Bangalore city railway station is about 2.5MLD (million litres per day) which is supplied by Bangalore Urban Water Supply & Sewerage Board (B.W.S.S.B). Out of the 2.5MLD, roughly 1/3rd is used for washing bogies. The general characteristics of composite wastewater sample before and after the treatment are:

Sl. No.	Parameters	Pre-treatment.	Post-treatment.
1	pH	7 - 8	7.6
2	Suspended Solids	300 – 350mg/l	13mg/l
3	Total Dissolved Solids	400mg/l	115mg/l
4	B.O.D @ 20° C	125mg/l	6.90mg/l
5	Chemical Oxygen Demand	320mg/l	117mg/l
6	Oil and Grease	12	

The total cost of the project was one Crore twenty lakhs rupees (2001). The approximate quantity of reusable water recovered from wastewater treatment was 600Kl/day (kilo litre per day). With a procurement cost of Rs. 73/ Kl from BWSSB, the project could save thirty-seven thousand seven hundred rupees per day with a yearly saving of Rs. 1, 31, 95,000/- or One crore thirty-two lakhs rupees approx. considering 350days of treatment.

Managing water demand through Evaporation and Seepage reduction
In 1967, arrangement for water supply to Bhavnagar city in Gujarat, India was made from Shetrunji Left Bank Canal of Shetrunji Dam

located at about 100Kms away. During the drought year of 1972-1973, it was observed that for getting every 5MGD of water at Bhavnagar, it was necessary to pump 32.5MGD of water in to the canal at Shetrunji Dam. The resultant loss including seepage at this unlined canal was found to be an astounding 85%. Consequently, a pipe line running 45Kms and costing Rs. 6crores was laid in 1977-79 to avert the loss due to seepage and loss due to evaporation.

The water supply to Ahmedabad city, Gujarat, India was fed from Dharoi reservoir some 130Kms from the city. The water pumped from the reservoir in to Sabarmati River, which was tapped through infiltration galleries located near the Ahmedabad city to cater to demand of the city residents. In February, 1987, it was found that even though the water was released at the rate of 120 cusecs from Dharoi, the surface discharge reduced to about 5 cusecs near the Ahmedabad city. In April during the same year, a discharge of 200cusecs from Dharoi reduced to about 12 cusecs of surface discharge near Ahmedabad, putting in question the wisdom of such an arrangement to bring water supply requirements for the city residents in such high loss scenarios.

In the above cases, the economic value of water conveyed is very high considering cost of pumping and conveying through series of canals. As such lining of canals is necessary. The lining of canals can reduce the seepage loss to 5 percent of their original value. The lining also reduces the maintenance cost of the canal by as much as 50 percent to 75 percent. The commonly used methods of lining are: (i) Hard Surface lining, (ii) Lining of exposed membranes, (iii) Buried membrane lining, (iv) Earth lining, (v) Use of soil sealants. Seepage losses are also highly pronounced in case of dams used for impounding water. In case of Masonry or Concrete dams, seepage losses can be controlled by one or more of the following techniques:

(i) Excavation and back filling with concrete in case of pervious or weak strata located at shallow depth or rock foundation.
(ii) Cement grouting to appropriate depths with or without admixtures.
(iii) Resin or epoxy grouting in problematic cases.

In case of Earthen dams, following techniques are more commonly used:
(i) Grouted cut off with clay cement, clay chemicals.
(ii) Partial rolled filled cut offs to appropriate limited depths supplemented by clay blankets upstream.
(iii) Cut offs of sheet piles driven to appropriate impervious strata or rocks.

Evaporation losses are high in tropical countries because of high temperature, arid conditions, and large number of sun shine days.

The factors affecting evaporation of water from an open water surface are:
- *Water surface area*: Evaporation depends on the surface area of water body exposed to atmosphere. More the surface area, more will be the evaporation, provided others conditions remain same.
- *Temperature*: The rate of emission of molecules from water is a function of temperature of the water body. Higher the temperature, greater is the evaporation rate.
- *Vapour pressure difference*: The rate at which molecules leave the surface depends on the vapour pressure of the liquid and rate at which molecules enter the water depends on the vapour pressure of the air. Evaporation depends on the difference of these two. Higher the difference, higher is the evaporation.
- *Wind effect*: Greater the veloecity of wind above water surface, greater will be the rate of evaporation.

Methods of determining evaporation: The various methods of determining the quantity or rate of evaporation from water surfaces are: *Water Budget/Storage Equation method*: The evaporation is measured by the following equation:

$$E = P + I - O + U \pm S,$$

Where, E = Evaporation, P = Precipitation, I = Surface inflow, O = Surface outflow, U = Underground inflow or outflow, S = Change in storage. S is (-) ve for any increase and (+) ve for any decrease in storage.

Limitations: (i) This requires accurate measurement of seepage losses. (ii) Equation may not hold good where large underground seepage occur. (iii) Equation does not hold good where large springs occur in lake beds.

Using Empirical Formulae: Rate of evaporation can be calculated from a number of empirical formulii.

Dalton's Formulae: Dalton was one of the first scientists to expound empirical formulae to measure the rate of evaporation.

$E = C (P_w - P_a)$, Where,

E = Rate of evaporation in inches per day of exposed surface.

P_w = Vapour pressure in the film of air next to water surface.

P_a = Vapour pressure in the air above the film.

C = Co-efficient dependent upon barometric pressure, wind veloecity, and other variable.

S.P.Ghosh & S.K.Sarkar of River Research Institute, West Bengal, India has developed the following equation for measuring Pan Evaporation:

$E = (1.3684 - 0.0189B) (0.41 + 0.136W) (e_s - e_d)$, Where,

E = Daily evaporation in inches.

B = Mean barometric pressure in inches of mercury.

W = Mean veloecity of ground wind in miles per hour.

e_s = Mean vapour pressure of saturated air at the temperature of water surface in inches of mercury.

e_d = Mean vapour pressure actually present in the air in inches of mercury.

The pan evaporation calculated by the above formulae can be converted to reservoir evaporation by multiplying with standard pan co-efficient. The accepted standard co-efficient for 1.22m (4') for US Class A Land Pan is 0.70.

Meyer's Formulae:

$E_L = K f(u) (e_w - e_a)$

E_L = Lake Evaporation (mm/day)

e_w = Saturation vapour pressure at water surface temperature (mm of mercury)

e_a = Actual vapour pressure of the overlying air at a specified height (mm of mercury)

f(u) = Wind pressure correction function

K = Co-efficient

(e_a is measured at the same height at which wind speed (u) is measured.

Measurement through Auxiliary pan: The water loss due to evaporation from standard pans are measured at regular intervals and multiplied by a suitable co-efficient to find the representative quantity of evaporation from a reservoir. Various types of pans used to measure evaporation quantity are:

USWB Class A Evaporation Pan: The USWB Class A Evaporation Pan is made of unpainted GI Sheet. The pan is kept on a wooden platform of height 15cm above ground to allow circulation of air below the pan. The depth of water on the pan is kept at 18cm to 20cm. The pan has a diameter of 1210mm and a height of 255mm.

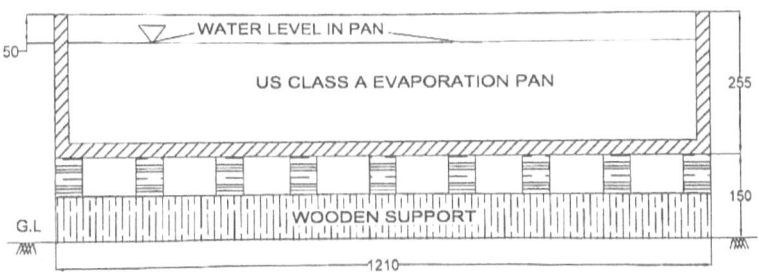

Evaporation Pan (ACAD representation: Author)

ISI Standard Pan: This Pan is made of copper sheet having a thickness of 0.9mm. It is painted white on the outside. To allow free circulation of air below the pan, the pan is placed on top of a square wooden platform of 1225mm x 1225mm in size and at height 100mm above the ground. The level of water is indicated by a gauge fitted on the pan. This pan has a diameter of 1220mm and a depth of 255mm. To protect the water in the pan it is covered with a hexagonal wire mesh net. This reduces the evaporation from this pan by about 14percent.

Colorado Sunken Pan: This pan is made up of unpainted GI sheet. It is of square shape having size of 920mm x 920mm. It is 460mm deep and buried in the ground within 100mm from the top.

USGS Floating Pan: The pan has a square size of 900mm and is 480mm deep. It is supported on drum floats in the middle of a raft of size 4.25m x 4.87m. The pan is kept afloat in a raft to simulate the large water body of the lake or reservoir, where evaporation is measured. The water level in the pan is kept at the same level as that of the lake, leaving a rim of 75mm. To reduce surges in the pan due to wave action, diagonal baffles are provided in the pan.

A pan used to measure evaporation is not true representative of evaporation in the lake. The pan of diameter of about 1.00m shows about 20% more than actual evaporation in the lake.

The height of rim of the pan acts as a barrier of wind action and effects the evaporation.

The rim casts shadows of different sizes on water surface.

Heat transfer characteristics of pan material are not similar to that of the reservoir where evaporation is measured.

For the reasons above, a Pan Coefficient is used, which when multiplied with the evaporation from the pan gives the evaporation in the lake.

Lake Evaporation = Pan Coefficient (Cp) x Pan Evaporation.

In general, average Pan Coefficients used for Class A Land Pan, ISI Pan (Modified Class A), Colorado Sunken Pan, USGS Floating Pan are 0.70, 0.80, 0.78, 0.80 respectively.

Various methods used in reducing evaporation losses are:
Providing wind breakers in regions of high wind veloecity: This method is suitable where surface area of reservoir is small as in case of ponds etc. The trees are planted in rows on the rims or banks of the water bodies. The tallest trees are planted in the middle and the smallest on the outer end to give a conical shape to tree lines. Spacing between plants varies from place to place depending on type of soil, climatic conditions etc. In general, following spacing are recommended: (a) Shrubs, 0.60m to 1.00m, (b) Medium height broad leaved trees, 1.50m to 2.00m, (c) Medium to Tall evergreen trees, and 2.10m to 2.40m, (d) Tall broad-leaved trees with conical crown, and 2.40m to 3.00m. Various types of vegetations have been prescribed by the Indian Council for Agricultural Research, New Delhi for different regions of India. However, this method

has limitations. The transpiration losses from trees themselves may be of considerable amount. In case of large water bodies this method is not effective as the inner water surface remains exposed to wind.

Providing physical cover over water bodies: This method is suitable for water reservoir with deep storages. Floating mats of appropriate materials are provided over the water surface. However, this method is costly and may not be cost effective in all situations. The covers reflect the energy input from the atmosphere and also act as a barrier and prevent escape of water vapour to outer atmosphere. Fixed covers are suitable for small water bodies. For large water bodies, floating covers are suitable. However, there is a danger of floating covers being swept over the spillways in case of large reservoirs.

Creating deeper water pockets to reduce surface area of exposure: By reducing the surface area and increasing the depth, the volume is maintained and at the same time area exposed evaporation reduces. The method was successfully employed in the Lake Worth, Texas, USA. In Gujarat, India, during the drought year of 1985-86, a shallow water pocket in Neyari 1 lake, covering an area of about 1.2 acre and containing some 1.4 mcft of water was pumped in to the main pool of water having a greater depth, thereby avoiding an evaporation loss of this 1.4 mcft of water and maintaining water supply to city of Rajkot.

Reducing surface water temperature: The surface water temperature is reduced by bubbling air through water body to the bottom and inducing water current, so that cooler water replaces the hot water on surface and thus reducing surface temperature and rate of evaporation.

Providing underground storages: The water surface exposure is avoided by storage of water in underground aquifers and cavities. The sub surface dams are constructed across a stream or rivulet. The ground water flow is restricted and the water level in the underground aquifers rises. Significant recharge of aquifer adjoining Talaji rivulet in Bhavnagar district of Gujarat, India was noticed with such an arrangement. Main advantage of such an arrangement is that submergence of valuable land and forest area is avoided.

Integrated Operation of Reservoirs: This system is followed in cases where there are number of reservoirs being used in an integrated way. The reservoirs are utilized in such a manner that the total exposed surface of water is kept minimum for the system as a whole during period of operations. Essentially, the shallow reservoirs with large exposed surface are utilized first. This method has been successfully used in case of Mumbai Municipal Corporation for their water supply scheme, Chennai Metropolitan Water Supply and Sewerage Board for Red Hills, Cholavaram, and Poondi reservoirs for water supply to Chennai city.

Using Evaporation retardants or monolayers: Chemicals such as Cetyl and Stearyl Alcohol, White spirit and Indicator oil are spread in form of powder or emulsion over the surface of water bodies through shore dispenser or dispenser mounted on rafts. The chemical emulsions form a blanket over the surface reducing evaporation rate by reflecting the energy inputs from the atmosphere. However, the chemicals forming the mono molecular layer over the water surface allows air to pass through it thereby making survival of aquatic life possible. Common chemicals used are:

Cetyl Alcohol (Hexadenacol) $C_{16}H_{33}OH$ Stearyl Alcohol (Octadenacol) $C_{18}H_{37}OH$ Ethoxylated Alcohols and Linear Alcohols, Linoxyd CS-40, Acilol TA 1618 (Cetyl Stearyl Alcohol)

An area of nine hundred thirty acres of Stephen Creek Reservoir in Australia was sprayed with Cetyl Alcohol. The reported reduction in loss of evaporation was thirty seven percent. In Umberumberka reservoir, Australia, a reduction of 50 percent of evaporation loss was reported for using such monolayers. In India, a reduction of as high as 48.2 percent in evaporation losses was reported in case of Baradari Tank in Maharastra. Experiments at following locations in India showed a saving of 30 percent or more in evaporation losses through use of Water Evapo-Retarders (WERs) or monolayers:

Sl.No.	Locations of Pond/Reservoir/ Tank.	Duration of experiment.	Name of chemical used	Percent saving in evaporation.
1	Badkhal Tank, Faridabad Haryana	Nov. 1958 to June, 1959 (8 months)	Cetyl Stearyl Alcohol (emulsion form)	37.80
2	Badkhal Tank, Faridabad Haryana	Nov. 1961 to June, 1962 (8 months)	Cetyl Stearyl Alcohol (emulsion form)	32.40

Sl.No.	Locations of Pond/Reservoir/ Tank.	Duration of experiment.	Name of chemical used	Percent saving in evaporation.
3	Takhat Sagar Tank, Jodhpur, Rajasthan.	June, 1964 to Oct. 1964 (5 months)	Cetyl Stearyl Alcohol (emulsion form)	48.20
4	Sagar Tanks, Ambar, Rajasthan	April-May, 1968	Cetyl Alcohol (powder form)	52.60
5	Indira Percolation Tank, Maharashtra	1977-1978	Linoxyd 0-22-40	37.00
6	Manchhu-1-Lake, Rajkot	Feb. 1986 to May, 1986	Cetyl Stearyl Alcohol (powder form)	35.00

MANAGEMENT OF WATER LOSSES IN WATER SUPPLY NETWORKS

In 1957, AWWA's (American Water Works Association) committee on 'Revenue Producing Water' drafted a report titled – "Revenue producing Vs Unaccounted for Water (UFW) to aid the water utilities in USA to evaluate their water supply system. The committee in its report put forward the concept of "metered ratio". The metered ratio is the ratio of metered sales to metered delivery to the distribution system. The difference between this ratio and 100 percent is the Unaccounted-for Water (UFW).

Statuses of unaccounted for water in some Asian cities/countries made available during an ADB (Asian Development Bank) organized seminar on "Control of Water Distribution System" for developing countries in 1983 at Singapore are as under:

Statistical Data of UFW for some developing countries in Asia (Year-1983)

Sl.No.	Name of City/Country	% UFW
1	Faisalabad (Pakistan)	50
2	Hyderabad (Pakistan)	30
3	Khulna (Bangladesh)	40
4	Dhaka (Bangladesh)	45
5	Greater Colombo (Sri Lanka)	30
6	Hadyai/Songhlia (Thailand)	50

Contd...

Sl.No.	Name of City/Country	% UFW
7	Bangkok (Thailand)	49
8	Metro Manila (Philippines)	51
9	Daegu City (Korea)	36.6
10	Malaysia	26.5
11	Hongkong	30

Data available for some cities in the USA suggests that between 2000 and 2010, the average leakages in their water systems stood at 12.3 percent. Some of the US cities and corresponding percentage of losses in water systems are as under:

City	Water loss in Water Supply Systems (%) between Year 2000 and 2010.	City	Water loss in Water Supply Systems (%) between Year 2000 and 2010.
Atlanta	31.4	Minneapolis	6.0
Boston	9.0	New York City	14.2
Charlotte	11.0	Orlando	10.0
Chicago	2.0	Philadelphia	26.5
Cleveland	28.7	Phoenix	6.6
Dallas	9.1	Pittsburg	26.0
Denver	5.0	Sacramento	10.0
Detroit	15.9	San Francisco	8.8
Houston	11.8	Seattle	8.0
Los Angeles	5.3	St.Louis	3.0
Miami	8.3	Washington D.C	14.4

Contributors of water leakages in water mains

Following factors are responsible for leakages in water mains:

(a) *High Pressure*: Flow of water in distribution mains is proportional to pressure inside the pipe. Higher the pressure, higher will be the wastage through leakages. An increase in pressure from 1.8 to 3.3 Kg/cm^2 results in approximately 30% increase in wastage. Wastage of water through various diameter of leakage holes are indicated as under:

Sl.No.	Size of leaking hole (mm)	Discharge in litres Under 3.3Kg/cm² per day.	Sl.No.	Size of leaking hole (mm)	Discharge in litres Under 3.3Kg/cm² per day.
1	0.4	1632	5	6.4	49032
2	0.8	2928	6	12.8	239712
3	1.6	17424	7	19.2	490320
4	3.2	20928	8	25.4	882576

(b) *Corrosive soil*: Corrosive soils cause corrosion to metal pipes vulnerable to corrosion. Soil corrosiveness is measured by its resistivity. Soil with resistivity 0-500 Ohm/m³ is considered alarming and detrimental to metal pipes vulnerable to corrosion.

(c) *Corrosive water*: Besides corrosive soils, corrosive water also affects pipes causing pin holes in pipe materials.

(d) *Poor quality of fittings & Bad workmanship*: Use of non-standard fittings may cause leakages, like wise poor workmanship involving inexperienced persons in laying process will give rise to leakage problems. Therefore, standardized pipe fitting confirming to laid down specifications. i.e. BIS etc. and involvement of expert and experienced personals in laying of distribution mains are advised.

(e) *Age of Service mains and communication pipes*: With aging, the metallic pipes especially GI pipes tend to corrode. At Gun metal ferrule connections, the GI pipe ends tend to corrode due to potential difference of these two metals.

(f) *Soil movements*: Newly laid pipes on soils, not properly compacted, are susceptible to discharge and leak due to soil shrinkages. The soil movement may also be due to landslides, unequal settlements, movement of soil mass due to earthquake etc.

(g) *Water hammer*: Sudden closure or opening of sluice valves will create pressure difference instantaneously. Sudden closure of a sluice valve will give rise to water hammer inside the pipe and may cause pipe bursts.

(h) *Effect of traffic*: Pipe networks are to be laid at least 1mtr below the ground surface to the top of the pipe to ensure adequate cushioning. Lower the cushioning over the pipe, greater is the tyre pressure transferred on the pipe and increased chance of pipe rupture. In general, tyre pressures of heavy vehicles which are at 6.00Kg/cm2 at road surface reduce to 0.30 Kg/cm^2 at a depth of 1.00 mtrs below the compacted road surface.

(i) *Damage by other utility services*: Often pipe lines are damaged by other utility services carried much later than when the pipes were laid like, telephone cables, electricity conduits or some drainage projects. Excavations conducted for such projects carried much later may damage the already existing pipe networks in absence of any knowledge regarding the alignments of such old existing pipe networks.

(j) *Leakages through gland packing of Sluice valves*: Leakages through gland packing of sluice valves are common problems. Daily use of sluice valves damages the gland packing which usually have a short life. The leakages are noticed when the seepages from the gland packing fill the valve chambers and over flow.

(k) *Leakages from cut off pipes*: Sometimes defaulting consumers' connections are cut due to many reasons such as illegally using pumps to draw water directly from the distribution mains, default in payment of water bills etc. When such a situation arises, the connection pipe is dislodged from the ferrule and the ferrule is closed or the connection pipe is cut and capped. Leakages occur in many occasions through this abandoned cut and capped pipes which go unnoticed for weeks in some cases and causing significant wastage of treated water.

Location of Leaks in pipe lines

The water passing through a ruptured pipe or a leak hole produces sound over a wide range of frequencies. This sound and the frequencies depend on (i) Water pressure inside the pipe (ii) Pipe material (iii) Size and Shape of leak or fracture (iv) Nature of ground surrounding the pipe where the water flows out. (v) Diameter of pipe.

Types of Sound: The water streaming out through a leakage in pipe produces sound energy which either dissipates along the pipe wall over a long distance or is absorbed very quickly in to the pipe wall. For metallic or hard walled pipe the sound transfer is extremely good and for non metallic or soft walled pipe, sound propagation is very poor. Some factors for producing good quality leak noise are – (i) High water pressure (ii) Hard back fill (iii) Small rupture (iv)Clean pipe (v) Metallic pipes (vi) Small diameter pipes. Some factors producing poor quality leak noise are – (i) Low water pressure (ii) Soft back fill (iii) Split mains (iv) Encrusted pipes (v) Soft/lined pipes (vi) Large diameter pipes.

Leak noise and pipe material: Water passing through ruptures or leak holes in a pipe creates pressure waves which are transmitted through body of water in pipe. The pressure waves spread sinusoidally in water on both sides. Everywhere these pressure waves meet the pipe material, cause mechanical oscillations. In case of metallic pipes, the frequency of these oscillations is high and usually ranges between 500Hz to 3000Hz and for non-metallic pipes, the frequency is low and ranges between 100 to 700 Hz. Some of the common pipe materials and their ranges of sound are as under:

Material	Diameter (mm)	Wall thickness(mm)	Frequency (kHz)	Range(m)
Steel	100	10	0.1	20000
Steel	100	10	1	5000
Steel	100	10	10	1250
Cast Iron	100	10	0.1	15000
Cast Iron	100	10	1	3000
Cast Iron	100	10	10	500
PVC	100	10	0.1	100
PVC	100	10	1	10
PVC	100	10	10	1

Source: Dr. Herbert Iann. SebaKMT, www.sebakmt.com

While locating leak noise frequency, three types of sound are considered-
(i) *Solid born vibration*: The oscillations of the pipe material are recorded through sensitive microphones. The frequency distribution of these signals range from 500Hz to 3000Hz.
(ii) *Ground noise signals*: The water escaping from the pipe under pressure impacts on the surrounding ground. The sound further moves up ward towards the surface in a funnel shape which is

detected by a ground microscope. The frequency distribution of these signals varies from 100Hz to 700Hz.

(iii) *Flow noise*: At pipe restrictions such as partially closed valves or at a pipe constriction, turbulent flows are created, giving rise to flow noise.

Three steps are involved in leak location:

(i) Zone measurement:
Zone measurement helps to identify the network parts where the losses are significant. The water distribution network is divided into zones comprising some 500 to 1000 connections which are metered or unmetered.

Stop Tap Method: This method involves testing a zone consisting of unmetered or partially metered consumers. The testing is carried out during non-supply hours. The water is supplied to a single feeder line through bye pass arrangements to supply water to the zone during non-supply hours for purpose of testing any leakages. All boundary valves and stop tapes at the consumer connections are closed before testing. The water supplied to the zone is measured by noting the reading on flow meter fixed on the feeder main after all the pipes are filled in the zone. The total quantity of water subsequently accepted by the zone during the test period is registered by the meter and is the wastage in the distribution system of the zone. The main advantages of this method are: (i) Statistical data in quantitative terms is known. (ii) Savings can be recorded in terms of liters of water saved. (iii) Gain in revenue to the water authority can be quantified.

Master Meter Method: A master meter is installed on the feeder main supplying water to a zone. All the boundary valves are closed, so that water coming to the zone through a single feeder main does not leave the zone. Supply readings are noted on the flow meter on the feeder pipe and consumption readings of all water meters are recorded before and at the end of the test period.

Quantity of Leakage = Flow meter quantity on the feeder pipe for zone supply during test period - ΣAll individual quantity recorded through water meters for all zone consumers during the test period. This method is suitable when all the consumers of a zone are metered.

The main advantages of this method are:
(a) More economical as compared to stop tap method.
(b) Does not require fixing of stop taps at consumer connections.
(c) No special supply is required.
(d) Less time required for testing.

(ii) Pre-location:

Pre-location with Leak Noise Correlator: When there is acoustic interference or when pipe network is inaccessible making the detection of ground noise difficult, Leak Noise Correlators can detect the precise position of a leak. The Leak Noise Correlator consists of (i) An internal radio receiver, (ii) A microprocessor to evaluate the data, (iii) Two highly sensitive piezo ceramic acceleration sensors, (iv) Two radio transmitters.

Correlator measures the distance to a leak from difference of time taken by a leak noise to travel to two pickups whose positions and relative distance between the two points are already known. From the distance between the pickups (D), and known veloecity of sound in the pipe material (V), the distance from a pickup (L) can thus be found

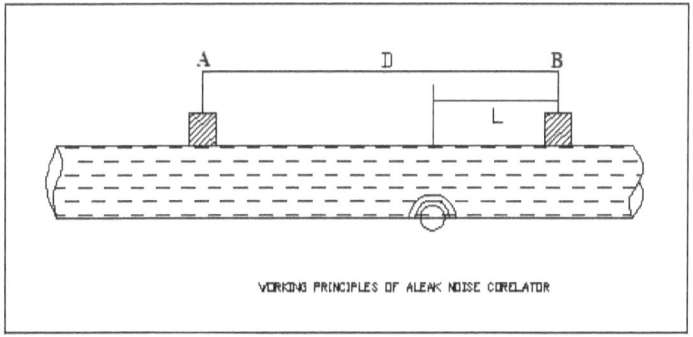

(ACAD representation: Author)

Time to reach pickup A = $(D - L)/V$
Time to reach pickup B = L/V
Difference of time (Td) = $((D - L)/V) - L/V$
The distance of leak from B, L = $(D - V*Td)/2$

At the leak location in a pipe, the water escaping from leak hole produces noise which moves on both sides of the leak towards 'A' and 'B'. The vibrations are picked at these points vide piezo sensors attached to pipe wall. These amplifier-transmitters send these signals to the base unit internal radio receiver. The microprocessor of correlator then calculates exact position of leak from the point 'B'. The different speeds of sound in various pipe materials are:

Speed of sound in different materials:

Material	Internal Diameter (mm)	Wall thickness(mm)	Speed of Sound (m/ms)
Steel	100	10	1.31
Steel	200	10	1.31
Cast Iron	200	10	1.28
Steel	200	20	1.02
Cast Iron	200	20	0.99
PVC	100	10	0.35

Pre-location with Noise Level Loggers: Leakages can be pre-located to confine the area of search to a certain areas or certain distances within which the chances of leakages are maximum and thus pin point leak location can be started in such particular areas or stretches avoiding any unnecessary excavation over a large stretch of pipe length. The noise loggers are installed occasionally or permanently on underground hydrants or on valve rods. The loggers come with transmitters and are placed at 200m to 400m for 3.5 bar pressure in case of metallic pipes and up to 100m in case of PVC pipes for same pressure. The automatic noise loggers can read signals at preset timings usually 2.00AM to 4.00AM at night. The leaks from a pipe burst create vibrations, which are transmitted via pipe and via water. These vibrations are stored in the noise loggers and signals are sent to a central 'commander' receiver. Usually 15 to 20 noise loggers are used at a time to transmit data to the 'commander'. Each logger is numbered and the data received from each noise logger is transferred from the commander to a PC, where print out can be taken for leak station, leak location, frequency, time etc. Through GPS positioning of these loggers, the noise loggers can be integrated on to a digital map of the distribution system. The stretch between noise loggers where maximum noise/vibration is recorded is selected for pin pointing of leakages.

(iii) Pin Pointing of leakage location:
After pre-locating the leakages in a pipe network, marking their exact location on the ground is done by (i) Visual Inspection, (ii) Sounding Rods, (iii) Tracer gas.

(i) *Visual Inspection*: Visual inspection of the ground surface along the pipe network is done in the pre-located area marked by a Leak noise correlator or a Noise level logger for any trace of leakages. For better results, regular visual surveys are needed covering entire zone area.

(ii) *Sounding Rods*: Sounding rods contain a 1.2m long hollow stick of mild steel having a diameter of 12mm and fitted with a brass cap of cup like shape of 5cm diameter on one end placed on the probable leak location along the pre-located positions, the other end of sounding rod is connected to a head phone for listening to any leak noise. For effective location, time used for survey is between 12.00AM to 4.00AM to avoid disturbances due to traffic noise. Well trained personnel can easily distinguish and detect a leak noise from other noises. Human ear can detect sound in the range of 20Hz to 20,000Hz at young age and this ability decreases with age. Modern leak detection equipment uses noise filter to filter external noises which makes it easy to conveniently detect a leak noise.

Electronic listening stick: To discard any possibility of error in listening, a listening stick with an electronic measurement device is used, where the sound signal is detected through an analogue needle display or through visual display of a histogram.

(iii) *Tracer gas*: Tracer gas is an inert gas and does not easily form compounds with other materials. It is soluble in water and is nontoxic and non-corrosive. The pipe line to be investigated is filled with dosed water to which tracer gas has been added under pressure. Higher the pressure, higher is the gas dissolved. The pressure of dosed water is maintained until it can be assumed that sufficient dosed water has leaked. As the pressure of the dosed water outside the main falls to atmospheric level, the gas comes out of the solution and permeates through the leakage moving vertically to reach the ground. Probe holes of 20cm depth are dug along the pipe line at a distance equal to depth of pipe invert from the ground. The tracer gas is then detected inserting detectors into these probe holes.

Authorities in U.K and Environmental Hygiene Department of the National Swedish Environment Protection Board have earlier accepted the use of Sulphur Hexafluoride (SF_6) as a tracer gas for leak detection in water distribution networks in their respective countries. A suitable concentration of standard mixture of 5 percent hydrogen and 95 percent nitrogen forming gas (95/5) is now used by many companies and water utilities in place of SF_6 which can be detected by a compound called H_2LUX. The mixture of oxygen and hydrogen with 5 percent hydrogen is classified as not flammable as per ISO 10156.

Infrastructure Leakage Index (ILI)

Infrastructure Leakage Index or ILI is a performance indicator of water supply systems which was introduced by A Lambert in 1999. This gives an alternative to traditional percentage method of finding real losses in water utilities around the world. The Infrastructure Leakage Index is the ratio of Current Annual Real Losses (CARL) (Ltrs/day) to Unavoidable Annual Real Losses (UARL)(Ltrs/day) in a water utility system.

$$ILI = CARL/UARL$$

Where, $UARL = (18*Lm + 0.80*Nc + 25*Lp)*P$
Where,
UARL = Unavoidable Annual Real Lossses
Lm = Length of Mains (Km)
Nc = Number of Service Connections (Main to Meter)
Lp = Length of Unmetered underground pipe from street edge to customer meter (Km)
P = Average Operating pressure at Average Zone Point (m)

An ILI of 2 should be set as a target limit for a well-managed system in developed country while an ILI of 5 is suggested for a more developing area (McKenzie and Lambert 2000). Liemberger (2005) has advocated two sets of ILI values, one set of limits (1>8) to be used for developed countries and another set (1>16) for developing countries.

ILI values around the world in 2007 (R.S McKenzie et al 2008)

Country	ILI	Country	ILI
Australia	0.5-1.7	South Africa	0.4-16.9
Austria	0.3-6.6	Thailand	46-543
Canada	1.1-9.0	U. K	1.7-1.8
Netherlands	0.3-0.6	USA	2.8-4.6
New Zealand	0.6-4.7		

ILI measured in areas of low-pressure systems were found to be of large values as seen in Bangkok water supply systems, where water pressure was below 5mtrs and the diameter of water mains, very large, usually of 3000mm. ILI datas are significant only when the water supply utility has a 24 hrs pressurized supply. ILI cannot be used as a performance indicator in case of intermittent water supply. For effective comparison of datas, different ranges of ILI values have to be considered for low pressure systems or intermittent water supply systems, as ILI measured in areas of intermittent water supply were found to be of very large values as in case of Asia, Africa and several parts of former Yugoslavia.

Water loss management in large diameter pipes

A leak detection programme carried by Gwinnett County, Georgia, USA on water supply networks of over 5400Kms spreading over a period of 12 months, detected over 500 leaks, resulting in water savings of 6.8MLD. The reports presented at 2007 AWWA DSS conference in Cleveland, Ohio, showed that the total volume of water saved in correcting these 534 leakages was 20.4gpm (gallon per minute) of which 14.3gpm was saved in 42 leaks detected in the water mains. In short, 7.8 percent of all leakages were responsible for 70 percent of all water lost.

In 2005, Southwest Florida Water Management District of US undertook leak detection programme wherein 735 leaks were located resulting in savings of 23.6 gpm in lost water. Of these 735 leaks, 69 numbers or 9.3 percent of leaks were found in water mains, which were responsible for 18.00gpm or 76 percent of water lost out of the 23.6 gpm of water lost through all of 735 leakages.

Often conventional leak detection methods tend to overlook leakages in large diameter pipes owing to various reasons such as, a greater number of leakages are found on the branches, the sluice valves, hydrants and other services. Also, it is often inconvenient to carry out survey in water mains in urban centres owing to heavy traffic conditions, greater depth of water mains below ground level etc. However new technologies like Tethered in Line Leak Location Systems provide greater accuracies in detection of leaks in large diameter pipes.

A CCTV camera is inserted into a water main through a 2" or larger diameter insertion nipple or tap. Tethered to the surface, the camera is gently pulled and flow along with water transmitting the real time video from inside the pipe line. The mobility of this camera is up to 6000ft or 1.8Kms in case of Sahara in Line Leak Detection Systems, a tethered in line type of leak detection system. Such a technology helps in:

(i) Locating lost line valves.
(ii) Investigating unexplained flow conditions.
(iii) Searching for illegal connections.
(iv) Locating and assessing tuberculation of all types inside the pipes.
(v) Visually inspecting pipe walls and liners.
(vi) Assessing internal corrosions of metallic pipes.
(vii) Inspecting a pipe near a known leak to help plan repairs.

Studies conducted over 1200Kms of water mains with Sahara in Line Leak Location Systems show an average of 1.2 leaks per kilometer of water mains. Average savings in lost water in some cities are-

Location	No of Leaks per mile	Savings in lost water.
Thames Water, London	3.2	40,000gpd
Dallas, Texas, USA	2.0	82,000gpd
Pennsylvania, USA	3.6	50,000gpd

CASE STUDIES IN WATER LOSS MANAGEMENT
Singapore Success Story:
The total land area of Singapore is approximately 700SqKm. The country's population in 1950 was about one million. There has been more than fourfold increase in Singapore population since 1950 which

has reached to about 4.5 million (2008). This has caused tremendous pressure on its water resources. The water demand of the country has increased from 142 MLD in 1950 to 1300 MLD in 2008, an increase of nearly 900 percent. However, Singapore's Public Utility Board (PUB), nodal agency for managing its water resources, has shown exemplary skills in balancing its supply and demand of water resources.

Singapore gets its water from four sources: (i) Water from local catchments, (ii) Imported water from neighboring Malaysia, (iii) Reclaimed water or NE water, (iv) Desalinated water. PUB, the national water authority of Singapore follows a two-pronged strategy, while on one hand it endeavors to augment its source of supply, on the other hand it tries to curb its water demand. PUB's conservation strategy includes (a) Price regulations, (b) Mandatory water conservation requirement and (c) People-Public-Private partnership in water conservation.

(a) Pricing: Public Utility Board (PUB) of Singapore uses pricing as an effective means for curbing water demand. In Singapore, water is considered an economic good and therefore pricing of water is done not only to recover fully the cost of production and supply but also to reflect the scarcity of this precious commodity. To encourage water conservation, a water conservation tax has been introduced in 1991. This water conservation tax has been made equal i.e. 30 percent for all categories of consumers since 1997.

Water Tariff in Singapore								
		Before 1ˢᵗ July'1997			From 1ˢᵗ July'2000		
Tariff Category	Con-sumption (m^3/month)	Tariff/m^3	Conservation Tax (in %)	Total/m^3	Tariff/m^3	Conservation Tax	Total/m^3
Domestic	1 to 20	56.0	0	56.0	117	30	152.1
	20 to 40	80.0	15	92.0	117	30	152.1
Non-Domestic	All units	117.0	20	140.4	117	30	152.1

(b) Mandatory conservation requirements: To deter wastage of water, legislative measures have been enacted to fine or prosecute in case of non-compliance with conservation measures taken by the government. Since 1983, water saving devices such as constant flow

regulators and self-closing delayed action taps was made mandatory in all non-domestic premises and in common areas of all high-rise residential apartments and condominiums. Since 1997, use of low capacity flushing cisterns (LCFC) that uses 4.5 liters /flush in place of normal 9.0 liters/flush was made mandatory for all residential premises, hotels, commercial buildings and industrial establishments.

(c) *People-Public-Private partnership initiative in water conservation*: Some of the major initiatives under this program are –
  (i) Holding National Save Water Campaigns every year, especially during the drier months to educate and aware citizens about conservation of water. These are done through (a) Publicity of conservation massages through media. (b) Distribution of Save Water Leaflets, stickers etc. (c) Setting up of Save Water exhibitions at constituency, hospitals, community clubs etc.

The "10-litre challenge": Another important step taken by PUB in 2006 was "10-litre Challenge". This initiative drawn in collaboration with Singapore Environment Council (SEC), a nongovernmental organization challenges every Singaporean to save at least 10 liters of water a day.

Water Efficient Building Program: The Public Utility Board (PUB) of Singapore runs a "Water efficient building program" to monitor usage pattern and conservation measures taken in commercial and industrial buildings. Under this program (i) PUB encourages the building owners to take ownership of their water consumption by emphasizing on water usage pattern. (ii) Encourages installation of water efficient appliances to avoid leakages and maintain flow rate of 2liters/min at all wash basin taps and 0.5liters/flush in all urinals. (iii) Building owners are also constantly encouraged to establish good water conservation practices.

Control of Unaccounted for Water (UFW): Unaccounted for water is the difference between quantity of water produced in water works as measured by meters and the quantity effectively reaching the consumers which is billed and accounted for. This UFW is due to leakages in specials, in pipe lines, in water retaining structures, illegal draw offs, apparent losses arising out of meter inaccuracies, and improper accounting in water used for cleaning reservoirs, cleaning and flushing washing mains, filling new mains etc. Between 1983 and 1993, PUB of

Singapore has carried out a ten-year replacement program where old CI and GI pipes have been replaced by cement lined DI pipes and other corrosion resistant pipes. All GI pipe connections at consumer points were replaced by copper or stainless-steel connections. Special emphasis was laid in stringent quality control in laying new water pipe lines so as to avoid any future leakages. As a result, the number of leakages reduced from 300 in 1985 to 31 in 2000. This could be achieved by constant vigil and public participation. It is reported that 92 percent of urgent complains regarding leakages are attended within 45 minutes. These tireless efforts of PUB, Singapore has resulted in reduction of its UFW from 10.6 percent in 1989 to 5 percent in 2008.

Water Loss Management Efforts in Manila

The water supply networks in Manila are supervised by Metropolitan Water Works and Sewerage System or MWSS. A massive increase in population during 1980s in Metro Manila and deteriorating water services conditions due to phenomenal percentage of non-revenue water (NRW) had put pressure on the Government to revamp MWSS in 1997. The historical records of NRW in Manila showed that, from 1986 through 1995, the water supply networks have incurred a percentage of NRW, 55% or more.

Year	%NRW	Year	%NRW
1986	66.4	1991	57.1
1987	59.7	1992	55.00
1988	57.7	1993	57.4
1989	57.7	1994	59.0
1990	57.7	1995	55.5

Historical records of NRW in Manila (Abelardo P. Basilio, Journal of IWWA).

In 1997, the MWSS was divided in to East and West zones with La Messa Treatment Plants – I & II falling in West Zone and Balara Treatment Plants – I & II under East zone. Both the zones were privatized in 1997. East Zone of MWSS has an approximate land area of 1400SqKm comprising 23 cities and municipal areas. In 1997, East zone had a total distribution network of 2400Kms but only 26% of the citizens there enjoyed 24hrs of water supply. The East zone was bid out and transferred to Manila Water Company (MWC) for up gradation and management of its water services including reduction in NRW.

Manila Water Company (MWC) initiatives: MWC divided the east zone water system in to 7 business areas with business area boundaries coinciding with hydraulic boundaries. Each business area was headed by a business area manager. Each business area focused in (i) Increased billed volume (ii) Reduction in None Revenue water (NRW) (iii) To promote community relation. MWC divided these business areas in to Demand Monitoring Zones (DMZ) or Territories headed by a Territory Business Manager (TBM). If a business area has 60000 to 100000 connections, a territory will comprise of 2000 to 5000 service connections. Each DMZ is further divided into District Monitoring Areas headed by one District Officer (DO). A DMA would normally have 500-1000 water service connections. Each TBM and DO is empowered to make important decisions and has the ownership and accountability of their own DMZs and DMAs respectively. These arrangements have ensured (i) On ground management more efficiently. (ii) Clear and better diagnosis of NRW reduction problems (iii) Quality execution of programs and projects. (iv) Efficient maintenance practices.

Corporate Program Coordinators (CPC): MWC assigned Corporate Program Coordinator to identify best practices followed in a DMZ or DMA and replicate the experience to other DMZs or DMAs. By end of 2007, MWC has established more than 1000 DMAs and CPCs were assigned to standardize the best practices followed.

Tubig Para Sa Barangay (TPSB): There was high incidence of NRW in areas of low-income communities due to illegal connections and damage of installations due illegal pilferages. MWC introduced water for depressed communities program called Tubig Para Sa Barangay (TPSB) and provided legalized water connections in these areas. By middle of 2008, Manila Water has completed over 644 TPSB projects in East Zone serving a total of 1.3 million poor people.

Pipe replacement: Asbestos Cement (AC) pressure pipes constituted 25 percent of the water mains in East Zone distribution networks prior to 1997 which contributed to high percentage of NRW due their failures. MWC conducted selective replacement of AC pipes as part of their initiatives. Installation of pressure reducing valves (PRVs) to adjust water pressure at night when there is lower water demand could prevent pipe burst and subsequently lowering NRW. The Manila Water Company's NRW reduction program has resulted in billed water volume of 1077MLD by June' 2008, a figure double the quantity in 1997. The Manila Water had 986000 households in their water consumer list compared to 214000 households in 1997. The MWC initiatives saw reduction in NRW from 63% in 1997 to 24% in 2007.

Managing apparent water losses through use of Unmeasured Flow Reducer (UFR)

Apparent losses are referred to as nonphysical losses, paper losses or commercial losses. Four main reasons attributed to apparent losses are:
(i) Meter accuracy error: Flow meter failing to register accurate flow in the pipe line.
(ii) Data Collection and transfer error: Erring meter reader collecting wrong readings from meter and Bill Clerk entering or transferring wrong readings in the bill registrar.
(iii) Data analysis error: Error in software or data entries.
(iv) Unauthorized consumption error: Pilferage of water through unauthorized connections from water pipe lines.

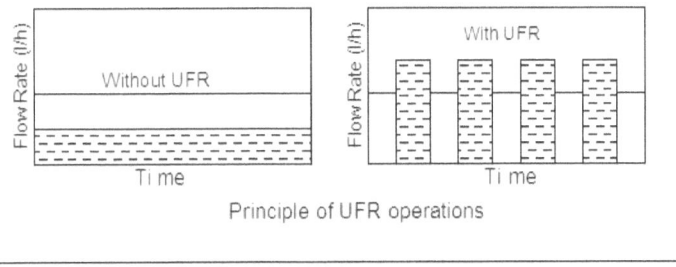

Principle of UFR operations

(ACAD representation: Author)

A domestic water meter installed improperly may create friction in its moving parts causing greater wear and tear and thereby making the water meter registrar lower than actual water flow through pipes. There is a threshold limit below which the meter cannot registrar any readings. With greater wear and tear the threshold limit also increases with passing of time. The flow meters are unable to registrar any readings at all if the flow rate is lower than this threshold limit. However, installation of Unmeasured Flow Reducers (UFRs) in conjunction with a water meter alters the flow pattern. When the flow rate is lower than the threshold limit, the UFR regulates the water flow to meter in such a way that there is no flow at all at certain time and in others, it is high enough to be registered by the meter. The water flow to the meter reaches in batches in such way that during period of availability, it is higher than the threshold limit and the flow meter can registrar any flow through the pipe/meter. Studies conducted in Israel in Jerusalem with use of some 360 UFRs. in one DMA and 120 UFRs. in another have shown that under registration by flow meters has actually gone down by 8.5percent and 9.9 percent respectively in both the district metered areas. Another study with use of UFRs. has shown a decrease of under registration of water flow through meters has gone down by 9.93 percent in Larnaka, Cyprus.

STRATEGIC MANAGEMENT OF WATER RESOURCES THROUGH VIRTUAL WATER TRADE

Majority of the food products require water for their production processes. The water used in the production process of an agricultural or industrial product is called the 'virtual water' embedded in that product. This water is 'virtual' because it is not physically contained anymore in the product. The real water content of a product is generally negligible if compared to the virtual water content. Example: To produce 1Kg of wheat in India, 1654 litres of water would be needed. Similarly, to produce 1Kg of Maize, 1937 litres of water is required. Livestock product generally requires more amount of water. To produce 1Kg of Cheese, it requires about 5000-7000 ltres of water. A 32 MB computer Chip weighing 2gm requires 3200 litres of water in its production.

The amount of water needed to produce these food products depend on the type of food product. Depending upon requirements of water for their production, food products are categorised as (i) Primary food products: e.g. Cereals, Vegetables etc. (ii) Processed foods: which are processed from primary food products, e.g. Oil, alcohol etc. from various primary food products. (iii) Transformed products: Animal products are considered as transformed products as these products mainly sustain on primary vegetable products, i.e. cereals, grass etc. (iv) By-products: These are the food products produced by crops grown for other purposes, e.g. cotton is grown primarily for production of fibre but its by-product cotton seed, is used in production of oil. (v) Multiple products: Coconut trees are grown and used in South Asian regions for multiple uses such as house building, raw material for production of sugar, coconut fruits, rope etc. Some animal products also give rise to several multiple products such as leather and fat for industry etc.

Therefore, virtual water content of a food product will depend on the types of the food product. The virtual water for a primary food product such as cereal or fruit is calculated by

$$VWV = ETa/Y$$

Where, Eta is the crop water requirement in m^3/ha.
And Y is the crop yield in Kg/ha.

It is estimated that, globally, $2m^3/cap/day$ of water is consumed for crop production. This excludes the water utilized by grass land.

The virtual water content of a live animal depends on the daily water needs of the animal, the water content of the food consumed during its life cycle up to that period and water used for cleaning of its habitat etc.

Virtual water content of different live animals for a few selected countries in m^3/ton of live animals are:

	Virtual water content of live animal (m3/ton)					
	Horses	Sheep	Goats	Bovine	Swine	Poultry
Australia	11707	6343	6585	11707	6117	2373
Canada	9619	5666	5440	9619	3268	1358
China	11186	5940	10016	11186	2160	3111
India	12729	6589	11237	12729	4175	8499
Ireland	7575	5246	4809	7575	2012	908
Italy	9581	5710	5407	9581	3459	1637
Japan	10751	5786	6105	10751	4325	2044
Netherlands	7680	5261	4823	7680	2086	914
Russian Fed	12310	6495	9055	12310	5488	4702
USA	10056	5715	5592	10056	3371	1304

(Chapagain& Hoekstra, Feb' 2003)

The virtual water content of agriculture or industrial or poultry products will differ according to the climatic condition and varying production process in a particular country. However, the global average virtual water content of few products is as under:

Sl.No.	Product	m^3/ton
1	Potato	160
2	Maize	450
3	Milk	900
4	Wheat	1200
5	Soybean	2300
6	Rice	2700
7	Poultry	2800
8	Eggs	4700
9	Cheese	5300
10	Pork	5900
11	Beef	16000

Nearly 5200Km^3 of water was embedded in food products in year 2000. The distribution of global water embedded in food products in the same year are:

Sl. No.	Food products	Percentage of water embedded
1	Alcohol	1.60
2	Fish & Sea food	8.00
3	Animal Products	15.60
4	Meat	29.60
5	Fruits	3.50
6	Vegetables	4.40
7	Oil	8.00
8	Sugar	5.50
9	Cereals	23.80
	Total	100.00

('Virtual water in food production and global trade'- Daniel Zimmer & Daniel Renault)

Approximately 1340 Km³ of virtual water was traded in food products in year 2000 globally. The quantity of virtual water in global virtual water trade in different food products in the same year is shown below:

Sl. No.	Food Products	Virtual Water Quantity in food traded. (Km³)
1	Alcohol	8.04
2	Fish & Sea food	191.62
3	Animal products	179.56
4	Meat	172.86
5	Fruits	32.16
6	Vegetables	22.78
7	Oil crops	176.88
8	Oil	201.00
9	Sugar	81.74
10	Cereals	272.02
	Total virtual water in food trade	1339.00

('Virtual water in food production and global trade'- Daniel Zimmer & Daniel Renault)

Thus, even by year 2000, the quantity of virtual water traded worldwide was 26 percent of the global virtual water budget in food products, with Oil & Oil crops contributing to 28 percent and Meat & Animal products together having a share of 26 percent.

Already many countries are adopting virtual water trade through import of water intensive products. Almost 60 to 90 percent of Jordan's domestic water is imported through trade in Virtual Water. It is estimated that the flow of virtual water to Middle East in form of grain imports are so large that it equals annual flow of river Nile. In the year 2000, Egypt saved as much as 5.8 billion litres of water by importing maize from other countries. This much quantity of water accounted for 10 percent of its national water allocation.

The water consumed for crop production and virtual water traded between continents in 1999 assuming an annual increase of one percent in water productivity in Km³/year is:

Sl. No.	Continent	Water for crop production	Virtual water imported	Virtual water exported	Net virtual water balance
1	North & Central America	684	164	317	-153
2	European Union	386	384	377	7
3	South America	445	52	175	-123
4	Asia	1673	426	182	244
5	Oceania	71	8	117	-109
6	Africa	241	97	19	78

China has significantly increased its net virtual water import from 7.02Km^3 in 1986 to 137.14Km^3 in 2009. Between 1986 and 2009, China imported 934Km^3 of virtual water more than its exported quantity at an average yearly rate of 39Km^3. All these imports & exports were through virtual water trade related to crops. 97 percent of all virtual water imports and 53 percent of all virtual water exports during 1986 to 2009 in China were through grain crops exports and imports of China. Before 2000, China mainly imported virtual water from North America but from year 2000 onwards it has increased its imports of virtual water from South America also. By 2009, China imported 64.95Km^3 of virtual water from South America also. U.S, Brazil, Argentina, Japan, Canada and South Korea are among the six countries that have traded more than 100Km^3 of virtual water with China between 1986 and 2009. While, US & Brazil each have exported more than 500Km^3 of virtual water to China, Japan & South Korea each have imported more than 100Km^3 of virtual water from China between 1986 to 2009. China has been a net exporter of Soybeans before 1996. However, with changes in trade policies, by 2003, the import of Soybeans by China exceeded the total domestic production of Soybeans in 2002. U.S, Brazil & Argentina together exported 886Km^3 of Soybean related virtual water to China between 1986 and 2009. Therefore, a well strategized virtual water trade policy could well minimize or stabilize a county's regional water imbalances within its national boundaries and contribute to water security.

It is estimated that virtual water content of international food trade in 2001 was 683Km³ /Yr. from point of view of exporting countries and producing the same quantity of food in importing countries would have required 1138 Km³/Yr. Thus 455 Km³/Yr. of global water was saved through virtual water trade in food products in 2001. Considering that the estimated total water used by crops in the world as 5400 Km³ in 2001, thus there was a saving of little more than 8 percent in global water budget in food in 2001.

Virtual water trade from India

Hoekstra & Hung estimated net virtual water export from India. As per their estimate, during the period from 1995 to 1999, India exported 191.8 Gm³ of virtual water and imported 19.5 Gm3 of virtual water, leading to net export of 172.3 Gm³ of virtual water from India. India ranked sixth among countries with net export during the same period. According to Chapagain and Hoekstra, India has exported 42.5 Gm³/yr of virtual water with net export of 25.4 Gm³/yr during the period 1997-2001. Out of the total export of 42.5 Gm³/yr., crop trade contributed 76 percent, livestock 8 percent, and industrial product 16 percent, whereas out of total import, 81 percent trade was related to crop, 2 percent to livestock and remaining 17 percent to the industrial products. Soybean was major crops for virtual export and palm oil, the major crops for virtual water import.

Thus, virtual water concept can be put to uses in two ways: (i) Virtual water can be seen as an alternative source of water and thus can be an instrument to achieve regional water security. Virtual water trade can be an instrument in solving geological problems and even prevent wars over water. (ii) The Virtual water content of a product tells something about the environmental impacts of consuming the product. Virtual water trade should contribute to local, national and regional food security, which requires not only appropriate trade agreements that respect a nation's right to decide on food security measures but also local distribution mechanisms ensuring access to food.

SOME GOOD PRACTICES IN WATER RESOURCES MANAGEMENT IN INDIA

(Source: Niti Ayog, Govt. of India)

(i) *Community managed drinking water supply system*: The Water and Sanitation Management Organisation (WASMO) in Gujarat, India brought together the communities in villages through Pani Samities, NGOs and few international Organisations like UNICEF, WASH along with World Bank etc. The villages covered by this drive were provided with piped water supply network and overhead tanks for distribution of drinking water supply to villages with participation of village communities. The technical assistance was provided by WASMO to ensure equitable share of drinking water to communities. Under these initiatives, 76.84 percent of rural households in Gujarat state were covered under piped water supply till 2014. Pani Samities were formed in 18,185 villages in the state. Water quality teams were formed in 16,860 villages, where 14,126 villages were supplied with field test kits to ascertain and ensure standard quality of water for drinking needs.

As a result of these initiatives, girls could continue education instead of fetching water from long distances, marked reduction in water borne diseases, overall improvement in health status in village communities and better living standards of the community were achieved. Engagement of communities made the program self-reliant and sustainable with minimum Govt. support.

(ii) *Jal Dal*: The Government School in Godawas in Rajasthan state, India experienced poor enrolment and attendance records. Children had to help their mothers in fetching water from far off places. Jal Bhagirathi Foundation with help of Gram Panchayat of the village constructed a 40(forty) thousand litres water tank in the school. The villagers created a Jal Sabha. Enlargement of the village pond was done for increasing the water storage quantity. For maintenance of the tank and for cleaning of roof etc, a 10 (ten) members student body was created, which was named Jal Dal. Jal Dal members were also responsible for cleaning of the silt chambers and smooth functioning of the hand pump inside school premise. The school children were involved in environment conservation drives and dissemination of information regarding water stress to villagers. The senior students started a 'piggy bank' by contributing one rupee every month and this fund was utilized for maintenance of the water tank and also for purchase of water during lean period of scanty rainfall. As a result of active participation of Jal Dal in water conservation, there has been marked reduction in school dropouts, increase in attendance, and decrease in water borne diseases, clean water being available throughout the year for entire village. The village became self-reliant and no longer dependent on outside supply through water tanks from private sources. The Jal Dal provided shining example of volunteerism and connectivity services. It also provided hands on experience on water saving methods through community participation.

(iii) *Mazhapolima Initiative*: People in rural Thrissur District in Kerala, India used to collect their drinking water from open dug wells used as individual sources. However, over usage of dug wells in the area led to drying up of most of these wells and deterioration of water quality in sources. Thrissur district administration in Kerala launched an artificial recharge program called Mazhapolima under the Thiruvilwama Gram Panchayat. Under this initiative, roof top rainwater harvesting systems were installed in the gram panchayat area using sand filters so that ground water is not contaminated. The intervention gave subsidies to poorer households in overexploited ground water blocks for installation of rain water harvesting systems. Employers of some 100NGOs received training for installation of RWH systems. As a result of this interventions 20,000 well recharging units were installed, over 100,000 (one lakh) people were benefitted, drinking water free from nitrates, iron content and reduced salinity could be made available. The amount spent on hiring water tankers was utilized for installation of rain water harvesting systems with aim of a sustainable water solution system for future needs.

(iv) *Adaptive Water Management*: The villagers of Mandli, Rajasthan, India had to collect water from far off places due to drying off of their main water source in a pond called Gawai Talab. The Gawai Talab had a capacity of 2869 Cubic Metres, which went dry during lean season due its small catchment and improper construction. Encouraged by success of Jal Sabhas in other regions, the villagers of Mandli with guidance from Jal Bhagirathi Foundation formed their own Jal Sabha. The members of the Jal Sabha took part in participatory planning exercise and generated funds through contribution of each household in the village. The villagers formed a Jal Kosh, where collected money would be placed so as to ensure maximum accountability with the collected fund in Jal Kosh. The area of Gawai Talab was increased from 2869 Cum to 5218 Cum. The money was utilized also for regular maintenance of the catchment by repair and renovation of water channels. Also, several trees were planted to improve greater water inflow. The pond has since been able to cater to need of drinking water in the village even in dry period and a long water crisis period was over, with availability of sweat drinking water throughout the year. The villagers could even supply water to neighbouring villages by introducing coupon system at a charge of Rs. 100 (One hundred) for a 4000 litres tanker.

(v) *Swajaldhara*: The Chinchojhar village of Dharampur Talukain Valsad, Gujarat had only one source of water in an open well, on which the entire village community depended for daily water needs. During dry season, this open well would often go dry, forcing the villagers, especially the women folk to walk up to 1.5Kms every day to reach another source in different habitation. Under Swajaldhara program, a collection tank was built near a spring at a height of 120 mtres from the village; the water so collected was brought through series of pipes and stored in a tank of ten thousand litres capacity constructed under the program at a height of 40 metres above the village. This ten thousand litres capacity tank acted as a source of water for stand posts, which were provided in the village for catering daily water needs of villagers. The water reaches these stand posts through gravity, saving electricity cost and cost in maintenance. Thus, villagers availed benefit though a simple intervention, where a perennial source of water was tapped and made use of, through principle of gravity and pressure to cater to the need of the tribal community.

(vi) *Mission Kakatiya*: It is a flagship program under Telengana state government in India. The main objective of the mission was to effectively utilize the 265TM of water allotted for minor irrigation sector in state from Godavari and Krishna river basins through enhancement of development in minor irrigation infrastructure and strengthening community based minor irrigation management through decentralization participation in management. Mission also aimed at restoration of tanks and water resources.

To achieve objectives, Gram Sabhas (village meetings) were called, where objectives and plans were discussed with villagers. Farmers were motivated to deposit silt for field preparation. Several district level coordination committees were formed. Steps like de-siltation of tanks, restoration of feeder channels, repair of bunds, weirs and sluices, raising of full tank level (FTL) in increased capacity where required, were carried out.

The mission intervention helped in increasing of storage capacity of tanks and other water bodies. It helped in making water available to small and medium farmers in particular. The water retention capacity of the sources increased, including increase in on farm moisture retention capacity. Mixing silt on farmland preparation reduced uses of chemical fertilizers, increased land water retention capacity. Also, there was marked improvement and rise in nutritive value of soil, resulting in diversification to high value crops and crop intensification. Other benefits were rise in ground water level in areas where intervention under mission Kakatiya was taken up. The increase in palm tree plantation on slopes added to income generation of rural people.

(vii) *Minor irrigation efforts at Narmada (Sanchore), Rajasthan, India*: This was an initiative taken by government of Rajasthan, India where minor irrigation technology was made mandatory at place of implementation in Narmada (Sanchore). Under the initiative, encouragement and enforcement of Participatory Irrigation Management (PIM) was taken up. Some 2236 Water Users Association (WUA) were formed for effective water management. Judicious usage of bio-drainage in command area and tree plantation along 1570 Kms length of canal was taken up. Salinity Resistant Crop plantation was proposed. Construction of canal network to utilize full irrigation potential was created. Conjunctive use ground and surface water was also taken up. As a result of these measures, the Culturable Command Area (CCA) increased from 1.35 hectares to 2.46 hectares with same quantity of water by adopting sprinkler irrigation system in entire project. There has been marked reduction of losses, both in cultivation and land loss. Irrigation and farm benefits were extended from 89 villages to 233 villages. Drinking water was also provided in 1541 villages and three towns, which were not included in project earlier. Increase in food production by 277 percent from 234 Crores to 1480 Crores based on year 2013-14 was achieved.

(viii) *Mulching*: Any organic material such as straw, saw dust, grass clippings, peat moss, leaves or paper etc that is spread on ground to protect the soil and roots of plants from effect of soil crusting, erosion, or freezing is known as mulch. For larger areas, the tilled soil acts as the mulch. Indian Cardamom Research Institute (ICRI), while studying and comparing soil fertility, has found that the humus content is more in areas of farm land where mulching was practiced, compared to neighbouring plantation. Application of mulches in Western Ghats in India showed that mulching resulted in several advantages: i. the plant growth is healthy. ii. Usually cardamom fields are replanted with new suckers in 8-10 years of cultivation, but mulching helped in retaining plantation for 17 years. Benefits of mulching: (i). Mulching helps in regulating temperature during hot summers and cold winters. (ii). Mulching helps to retain soil moisture. (iii). It helps in nitrgen fixation. iv. Growth of unwanted weeds in farm land can be controlled through mulching technique.

(ix) *Minor irrigation in Gujarat*: A special purpose vehicle called Gujarat Green Revolution Company was formed in the state of Gujarat, India to promote and implement micro-irrigation schemes in the state. The intervention intended to bring an integrated approach in farming methods and promote uniformity under various schemes. Farmers were educated in water management techniques and benefits of value addition in crop production and marketing their products. The initiative educated and encouraged farmers in storage of water through Jal Sanchay Abhiyan, in which micro-irrigation scheme was an integral component. This intervention brought a green revolution in the state of Gujarat. By 2014, since implementation of the project, 640853 farmers in a total area of 1034930 hactares adopted micro-irrigation system (MIS) and reaped benefit of the program. Out of the 1034930 hectares, 496305 hectares was covered under drip-irrigation and 538625 hectares under sprinkler irrigation. This has brought a second green revolution in the state. 131293 tribal farmers in cumulative area of 178745 hectares benefitted through the project, giving rise in their source of agricultural income. Effective storage and management of water were key to the success.

(x) *Bhungroo*: "Bhungroo" which means a hollow pipe or straw in Gujarati language is a water management system that stores rain water during monsoon or rainy season to be used for irrigation later. Essentially, it contains a PVC pipe with slotted holes which is inserted in the ground for a depth ranging from 20-70 metres. The site housing the pipe is dug up to 3 mtrs depth in a 2.5x2.5 mtrs square pit, which is filled with sand gravel that acts as filter for rain water. The raw water enters through the slots in the pipe and replenishes the ground water table. The Bhungroo system was implemented by government of Gujarat in India, which was carried out at sites identified by gram panchayats, design and estimates of which was done under "Mahatma Gandhi National Rural Employment Guarantee Scheme" (MGNREGS). The farmers were trained in installation of Bhungroos. The system installation resulted in enabling communities to farming for more than half of the year. The system helped to bring down the salinity of ground water making it fit for agricultural use. The system can store up to 40 million litres of water, and can supply stored water for as long as seven months. The use of Bhungroo in pilot project in Gujarat provided food security and sustainable livelihoods to more than 18000 marginal farmers (with over 96000 dependent families). Thus, system has reduced the hardship of women folk, making them the chief owner and expert in this field.

(xi) *Pani Panchayat*: Department of water resources, Orissa under Orissa Water Resources Consolidation Project (OWRCP) introduced participatory irrigation management (PIM) on a pilot basis in 1995 under the banner of Farmers Organization and Turnover (FOT). After experiencing its overall success and achievements, the project was extended to cover all the commands under major, medium, minor and lift irrigation projects. The main objectives were: i. To secure equitable distribution of water among its users. ii. Proper maintenance of irrigation system. iii. Efficient utilization of water for optimized agricultural production. iv. To protect environment to ensure ecological balance. v. inculcating sense of ownership of the irrigation system among farmers. To achieve these goals, "The Odisha Pani Panchayat Act 2002" was introduced in November'2002 & "The Pani Panchayat Rules, 2003" was introduced in April'2003. Before these acts and rules came into effect, Pani Panchayats were registered under Society Registration Act, 1860. After introduction of these acts and rules, several reforms took place, like – i. Inclusion of fisher folks in Pani Panchayats, ii. Increase in tenure of office bearers and executive body from three years to six years, subject to replacement of fifty percent by lottery, iii. Election of members on rotational basis, iv. Inclusion of women by reservation of one-third of the total number of seats in executive committee; "Pani Panchayats Samachar" was introduced as a quarterly publication to help in exchange of knowledge. Best performing Pani Panchayats were awarded during yearly Pani Panchayat Fortnight being celebrated from 26thDecemeber to 9th January every year. Regular on and off campus training programs are undertaken by Water and Law Management Institute (WALMI), Orissa. These efforts helped in sustained and effective management of water resources.

(xii) *System of Water for Agriculture Rejuvination (SWAR)*: This system highlights the importance of root zone as an ecosystem that retains soil micro-organism besides rationing plant water requirement. SWAR was introduced in Hyderabad by Centre for Environment Concerns (CEC). Under this initiative, water is stored in overhead tanks, from where it is passed through small diameter pipes to locally made clay pots which are buried near the root area. The water from these clay pots are transmitted through micro tubes which are connected to clay pots which passes through sand pouches, that makes available enough misture in root zone for healthy growth of plants. The micro tubes ensure slow oozing of water and proper rationing and provide water for longer durations. Thus, SWAR uses very low quantity of water and there is no wastage of water. The system uses one-eighth of water compared to drip irrigation. It showed immediate results in terms of plant health and farmers' income. Due to these promising results, SWAR received 'the global Champion Innovation prize for Water and Forestry' at 'Paris International Agricultural Show' in 2015. This innovation has been well received by government of Andhra Pradesh, India, which has placed order for implementation of this system in 400 acres of land in Anantpur, Krnool, and Chittoor.

(xiv) *Orange city water project*: The water consumers of Nagpur Municipal Corporation (NMC) received intermittent supply for 8-10 hours a day. Due to dysfunctional and non-existent meters, only 175 MLD out of total 575 MLD of treated water could be billed and revenue collected. Erratic supply of water directly benefited private suppliers, who were supplying water through tankers for hefty prices. To solve these problems and increase revenue earnings, NMC passed a resolution for 24x7 water supply in Nagpur city. A consortium of Veolia and Vishwaraj was formed through transparent bidding and given responsibility. Under the terms of agreement, the asset ownership lies with NMC, operations are to be looked after by Veolia-Vishwaraj Consortium, which formed 'Orange City Water Pvt. Ltd'. All the revenues are collected by the consortium. The consortium operates and maintains the water networks under an Operation and Maintenance contract for 25 Years. NMC augmented the treatment capacity at the Anhar River from 120MLD to 240MLD on its part; OCWPL took over the city water supply and replaced 85000 out of 321000 connections along 450kms long pipe network. Nearly 100,000 unauthorised connections were identified and regularised. Service delivery was improved by augmentation of infrastructure and increase in overhead reservoir capacities. All bill payments are managed through zonal level kiosks set up by OCWPL. Round the clock call centres was set up to meet public grievances related to water supply. The intervention helped in increase of revenue, better efficiency in supply, lower loses and greater satisfaction to water consumers in terms of delivery in quantity and quality of water through PPP mode.

(xv) *Non-revenue cell of SMC*: It became a cause of concern, when non-revenue water (NRW) score received by Surat Municipal Corporation (SMC) was a dismal 'D'. The water network under SMC suffered from leakages and breakdown in the system. There were complaints about water pressure in the system from consumers at different zones. Some leakage mappings and repairs were undertaken by SMC at various zones, however all these were on piece meal basis. To solve these problems in a comprehensive manner, NRW cell was set up for thorough estimation of water losses in the system and to keep NRW within 20 percent, water audit and leakage mappings exercise was conducted in city's core areas. Leakage mappings took cognizance of historical complaints of consumers to identify location and status. Based on leakage mappings and ground assessment, defective pipe lines were removed stage wise and extended zone wise from core areas. As a part of leakage repair exercise, leakages in pipe size greater than 750mm was out sourced to private operators, leakages in pipes between diameter 550-750mm was out sourced to Ahmadabad Municipal Corporation and leakages below 550mm diameter was repaired by zonal offices under SMC. As a result of this exercise, number of leakages per kilometre of pipe line reduced dramatically during 2011 to 2017. The length of water supply network increased to 3000 kilometres. Consumer complaints reduced to large extent. There was considerable water savings. Thus creation of dedicated institutional mechanism like NRW cells saw positive enthusiasm and satisfaction of consumers.

(xvi) *Birkha Bawri*: A private township called The Umaid Heritage site located at south east of Umaid Bhawan Palace in Rajasthan, India were facing hardships due to scarce water supply and also faced problems of water logging during rainy seasons. Umaid Heritage Real Estate agency developed the Birkha Bawari structure, which draws its inspiration from traditional stepped well structure for conservation of rainwater. The structure was developed to collect rainwater on its longitudinal direction from catchment on both sides. Being a step well of about 18 mtrs depth, it could collect and store up to 17.5 million litres of harvested rain water annually. Due to its depth, it received shade from the sides which lowered evaporation losses. The water was collected from roof tops of housing colony through a series of pipes and also through drainage conduits on its longitudinal edges, through natural slopes. The intervention helped in reduction of ground water extraction in the area by as much as 50 percent. Residents received water during lean period of scanty rainfall. The intervention also reduced long time water logging problem during rainy season reducing load on the existing storm water drainage system. Reduction in dependency on water tankers saved residents 2.36 Crores of rupees annually on water costs on private sources. All these, have resulted in a sustainable water management solution and also increased the property value in the area manifold.

Water Resources Management through Legislation

Water law is the area of law dealing with ownership, access and control of water.

It is also concerned with interstate and Trans boundary dimensions of water, divisions of power between the central government, states, local bodies (such as municipalities in urban areas and panchayats in rural areas), the public and private organization, as well as issues of water quality and health & environmental implications.

Water Laws in India:
1. There is no comprehensive water legislation in India.
2. Constitution of India gives power to the state government to enact water law.

Objectives of water law
1. Allocation of water for different uses
2. Setting up of priorities among different uses of water
3. Conservation of water resources
4. Implementation of fundamental human rights to water
5. Maintain quality of water sufficient for its various uses
6. Ensure water for human survival and poverty eradication.

Fundamental Human Right to Water

Over the years, existing framework concerning water has been complimented by human right dimensions. The core message is that all human beings are entitled to equal and non-discriminatory supply of a sufficient amount of water. This has led to demand for legal recognition of right to water and corresponding policy changes and conditions across the globe. In India, fundamental right to water has been confirmed by the courts.

In some cases, courts have made it clear that the government has an obligation to provide water.

In case of "Subhas Kumar Vs State of Bihar (1991)", Supreme Court of India ruled: the right to life includes "the right of enjoyment of pollution free water and air for full enjoyment of life" (Para - 7)

In case of "Narmada Bachao Andolan Vs Union of India (2000), Supreme Court of India": water is the basic need for the survival of the human beings and is part of right of life and human rights as enshrined in articles 21 of the Constitution of India (Para - 244)

In case of "Shajiur Joseph Vs State of Kerala (2006)": High Court ruled: the government is bound to provide drinking water to the public and that this should be the foremost duty of the government. Additionally, the judges ruled that the failure of the state to provide safe drinking water to citizens amounted to a violation of article 21 of the Constitution (Para 3)

National Rural Drinking Water Programme (NRDWP) and National Water Policy 2002, calls water a 'basic need' or 'a basic human need' rather than a basic human right (Para 1.1)

Some countries have gone further than India in pronouncing access to water as a fundamental human right
1. South Africa: Right to have access to sufficient water
2. Uruguay: Access to potable water and access to sanitation

Right to water in International Document
1. General comment (no 15) on the right to water adopted by United Nations committee on Economic Social and Cultural Rights in 2002 states that the human right to water "entitles everyone to sufficient, safe, acceptable physically accessible affordable water for domestic uses".
2. Convention on Elimination of Discrimination Against women, 1979. Article 14(2) (h) explicitly mentions about provision of water and sanitations to women
3. Convention on the Right of the child, 1989, Article 24(2) (c) mentions right to safe drinking water of a child from a non-polluted source.
4. Draft guidelines for the realisation of the Right to Drinking Water and Sanitation 2005, adopted by United Nations Sub Commission on the Promotion and Protection of Human Rights says that "Everyone has a right to a sufficient quantity of clear water for personal and domestic uses" (Para 1.1)

Supreme Court of India used a legal construct known as public trust which states that government is not an owner but a trustee of water and as much responsible for protecting and preserving water for on behalf of the beneficiary or the public.

This requires the government to manage and develop water without depriving any individual or group from accessing or significantly affecting ecosystem needs. As a result, neither the government nor individuals can exercise absolute rights over water.

Public trust was more pronounced in the case, "M.C Mehta Vs Kalmal Nath 1997", where the court ruling was based on public trust doctrine. The court concluded in the Span Motels encroachment case to turn course of the river Beas: "Our legal system based on English common law includes the public trust doctrine as a part of its jurisprudence. The state is the trustee of all-natural resources which are by nature meant for public use and enjoyment. Public at large is the beneficiary of the sea shore, running water, airs, forests and ecologically fragile lands. The state as a trustee is under a legal duty to protect the natural resources. These resources meant for public use cannot be converted in to private ownership" (Para 34)

Water as an economic good:
One of the influential aspects of water sector law reforms is the principle that, water is now being viewed as an economic good. This principle considers water as a commodity.

This has risen from fact that earlier water was considered as a free unlimited natural resource, which was collected through infrastructure constructed by government and supplied to public for use in agricultural or for domestic uses. This has given rise to unscrupulous use of natural resources and wastage of highest order.

The reforms in water policies in government now envisage water as a commodity. Which means it can be used as an economic good and can be traded. Since it has price attached to it, there will be prefixed quantity that an individual can use. If a person is not using the entire quantity, he can sell his unused quantity. This way, a rural person can sell his unused water to an industry and earn profit.

Thus, rights of water can be transferred from one person to another person for money. This view was adopted in Swajaldhara Program and World Bank assisted projects for rural drinking water supply reforms.

Ministry of Rural Development spearheaded the introduction of Swajaldhara through adoption of Swajaldhara guidelines. The main features of Swajaldhara guidelines were:

1. The programme adopts community participation based on empowerment of villagers to ensure their full participation in choice of drinking water scheme, planning, design, implementations, control of finances and management.
2. Ownership of the drinking water assets is handed over to the Panchayat.
3. The Panchayats/communities have the powers to plan, implement, operate, maintain and manage all water supply and sanitation scheme.
4. The community is required to contribute 10 percent of the capital cost in form of cash/kind/ labour/ land as a combination to benefit the scheme.
5. The central government provides 90 percent of the project cost as grant.
6. The beneficiaries are 100 percent responsible for operation, maintenance and management costs of the schemes constructed under the programme. All repair costs, salary costs, electricity are born by users.

Law relating to Urban Water Supply (UWS) in India: Urban Water Supply comes within the purview of state Governments. State Governments are authorised to make laws on UWS, Central Government has minimal role. However, state government is not directly involved, instead it is the domain of Urban Local Bodies (ULBs) such as municipalities and corporations.

1. Most municipal acts contain a chapter on 'Water Supply & Sanitation' which make water supply & sanitation an obligatory function of these local bodies. Judiciary in India also endorsed this duty of local bodies in some cases.

2. Some states have adopted law to constitute separate authority for water supply & sanitation. Example:
   a. Kerala Water Supply Sewerage Act 1986, b. Karnataka Urban Water Supply and Sewerage Board Act 1973, c. U.P Water Supply & Sewerage Act 1975, d. Assam Urban Water Supply & Sewerage Board Act 1985.

Model Municipal Laws, drafted by the Ministry of Urban Development & Poverty Alleviation state under Para 47 (1) that every municipality shall:
a. provide on its own or arrange to provide through any agency the following core municipal services: i. water supply for domestic, industrial and commercial purposes.

Para 167: not withstanding anything contained elsewhere in this Act, but subject to the provisions of any state law relating to planning, development, operation, maintenance and management of municipal infrastructures and services, Municipality may in the discharge of its functions specified under section 47, section 48 and section 49 -
a. Promote the undertaking of any project for supply of urban environmental infrastructures or services by participation of a company, firm, society, trust on any body corporate or any institution, or Government agency or any agency under any other law for the time being in operations of such project of a Municipality irrespective of its cost.

Irrigation Laws

Most of the surface and ground water in India is used for irrigation. Therefore, irrigation laws are of prime importance in regulation of water for such usage

Many changes have been introduced in irrigation laws to give importance and precedence to Participatory Irrigation Management (PIM), where in greater emphasis is given to user's participation in irrigation management. Thus, there has been a significant shift in perception & practices where earlier the Government does all works of construction and maintenance of irrigation canals and water allocation to farmers.

Some states like Andhra Pradesh, Rajasthan, Orissa, Madhya Pradesh, TamilNadu, Maharashtra and Chattisgarh have already passed specific Farmers Participation in Management of Irrigation System laws where in transfer of some responsibilities of irrigation management from Government agencies to Water Users Associate (WUA) at primary level has been enumerated.

Some salient features of Water Users Association laws:
1. There is a three-tier structure for WUA farmers' organization. The WUA or Pani Panchayat, the distributor committee and project committee.
2. The criteria for deciding the boundary of the command area as a water user area is based on hydraulic basis.
3. A WUA is established for every water user area and it consists of all water users who are land owners in the area as members.
4. A WUA is meant to be governed and controlled by people that both pay for the services the association offers and receive benefits.
5. WUAs will benefit from a more assured water supply and more control over water allotted to them. They also have the right to use ground water in their command area on top of the entitlement they receive from canals.
6. Functions of WUAs are to regulate and monitor water distribution among WUA members, assessment of member's share, responsibility of supply equitable proportions of water of its members, collection of service charges, carrying out maintenances and repairs of canal systems and resolution of disputes among members.

As of 2007, there were 55,501 WUAs in India covering an area of 10,230,000 hectares in India. A 2007 Planning Commission (now Niti Ayog) report states that 58 percent of the irrigation area is sourced from ground water and 80 percent of water need in country is sourced from ground water. Due to this, there is wide spread depletion of ground water resources. A study conducted by Central Ground Water Board found that out of 5723 assessment units 226 units are found to be critical and 839 assessment units were found over exploited. Therefore, there was urgent need for legal regulation to prevent ground water exploitation.

The common law principle which is still a part of ground water law in country considered ground water as part and parcel of the land. The legal implications of common law is that the owner of the land could dig wells in his land and extract as much water as he can or wants. The land owner was not responsible for any damage caused to water resources of his neighbour on account of his over extraction on his land. This legal principle could be seen in some laws dealing water land rights, i.e. "Indian Easement Act' 1886". This principle was also endorsed by courts during pre-independence era.

Common law principles have also been applied in post independent India, for example in Plachimada-Coca Cola case, where the division bench in Kerala high court asserted the primacy of land owner's control over ground water in the absence of a specific law prohibiting extraction.

Water legislations focus on national level in India:
1. Water legislation at national level is aimed only at preventing pollution of water and its various resources,
2. Current legislations don't cover the area of conservation of water as a resources and usage in an efficient manner.
3. There is no legal framework to deal with water use efficiency (WUE)
4. Existing Acts on water at national level:
a. The Water (Prevention and Control of Pollution) Act 1974 (amended 1988,1992)
b. The Water (Prevention and Control of Pollution) Cess Act 1977(amended 2003)
c. EPA (Environmental Pollution Act)

The Water Cess Act was introduced to augment the resources of central & state board constituted under the Water (Prevention and Control of Pollution) Act 1974 by levy and collection of cess on water consumed by certain industries and by local bodies/ authorities.

The Water Cess Act also mandates that all industries and local bodies would install water meter to ascertain quantity of water being used by them and to levy cess based on water usage by them.

The rate of cess (being very less) is linked with fulfillment of condition related to effluent standards and not linked to quantity used by industries or local authorities.

The objective of the Water Cess Act is to generate resources for institutional framework to create, to enforce water pollution laws and not to acknowledge that water is a scarce resource.

The Guwahati Water Bodies (Preservation and Conservation) Act 2008:
This act of Government of Assam, India received assent of the Governor on 5th August 2008, and was passed to provide for preservation, protection, conservation, regulation and maintenance of natural water reservoir and convert in to eco-tourism recreation centre to suit the ecological balance within the jurisdiction of Guwahati Metropolitan Development Authority and to protect the water bodies from encroachers and damages.

Andhra Pradesh Water Resources Regulatory Commission Act 2009:
This act was brought to provide for the establishment of the Andhra Pradesh Water Resources Regulatory Commission for regulation of water resources within the state of Andhra Pradesh to facilitate effective utilization of water resources within the state to ensure its sustainable and scientific management for drinking, agriculture, industrial and other purposes and matters connected therewith or incidental thereto.

The commission has following powers, functions and duties among others:-

1. To determine the water requirement of various categories of users (such as irrigation, municipal, rural drinking water, industry etc) on a yearly/season basis.
2. To determine the adequate operations and maintenance cost (O&M) of irrigation/ multipurpose water projects. The state shall ensure provisions for full operation and maintenance requirements of such projects as determined by the commission, through an appropriate budgetary support.
3. To promote efficient management of irrigation water

4. To promote efficient use of water resources and minimizing wastage of water by (a) Fixing and monitoring implementation of stipulated quality standards for management of water resources by various water users/departments and recommend actions against violators.
5. Fixing and monitoring implementation of stipulated quality standards for disposal of waste water by various water users and recommend actions against violations.
6. Ensure publication of Annual Report containing water audit, irrigation status etc.

Andhra Pradesh water land and Trees Act 2002:
This Act was passed to promote water conservation and tree cover and regulate the exploitations and use of ground and surface water for protection and conservation of water sources, land and environment and matters connected therewith or incidental thereto.

Under the Act, the state Government shall, by notification constitute an authority called "Andhra Pradesh State Water, Land and Trees Authority".

Salient features of the Act:
All ground water resources in the state shall be regulated by this authority. Some regulations of the authority require -
1. Registration of wells: All well owners to registrar their wells with authority (Para 8(1))
2. The designated officer, with approval of the Authority, may prohibit water pumping by individuals, group of individuals or private organization in any particular area, if in his view such water pumping in such area is likely to cause damage of the level of ground water or cause deterioration or damage to natural resources or environment, for a period of not more than six months, which after review may be extended for a further period of not more than six months at a time. (Para 9 (1)).
3. The Act empowers the Authority to regulate sinking of any private wells near source of public drinking water (Para 10(1)).
4. Authority may, on the advice of the technical officers, that any existing well is found to be adversely affecting any public drinking

water source, after giving the owner, a reasonable opportunity of being heard by an order, prohibit the extraction of water for commercial, industrial, irrigation or any other purposes from such well for a period of not more than six months, which after review may be extended for not more than six months at a time (Para 12(1))
5. To improve the ground water resource, by harvesting and recharge, the authority may issue guidelines for constructing appropriate water harvesting structures in all residential, commercial and other premises and open spaces having an area of not less than 200 square meters (Para 17, 1))
6. In irrigation command areas, water user association shall ensure optimum use of surface and ground water and for this purpose, the water users associations shall adopt the measures suggested by the designated officer (Para 22)
7. The authority may notify water bodies like, village ponds and minor irrigation tanks along with nalas (water course or drainage course) as heritage bodies and conservation areas to prevent conversions of their intended use and the authority shall take all measures to permanently demarcate the boundaries through the department of the Government or the organisation concerned as per the memoirs of lakes/ tanks/ponds/nalas (water course or drainage course) and shall take measures to evict and prevent encroachment (Para 23(1)).

Assam Irrigation Act, 1983:
This act though made in 1983, received assent of the President in 1989 and then published in the Assam Gazatte extraordinary no 68, dtd 19-05-1989. This is an Act to provide for the application and use of water for purposes of irrigation.

Some highlights of this Act are:
Notification of intention to apply or regulate water for irrigation:
1. "Whenever it appears expedient to the State Government that the water of any river (including tributary) or a stream flowing in a natural channel or of any lake or any other natural collection of still water or ground water or water flowing in a channel where such water or

part thereof, is received from an irrigation work constructed by the state government, whether by percolation, regeneration, release or otherwise, should be applied or used for the purpose of any existing or projected irrigation work or for the regulation, supply or storage of water, the state government may, by notification in the manner prescribed, express its intention to that effect specifying therein the land or block of land to which said water will be so applied or used. (Para 3)
2. The Divisional Irrigation officer (Executive Engineer), either on his own motion or on the application of any one or more of the owners or occupiers of any land within the culturable command area of any irrigation work, may prepare a supply scheme to provide for the supply and distribution of water from an irrigation work through the supply works to the land within the culturable command area of an irrigation work (Para -16)
3. Owner or occupiers to maintain supply works properly:
   a. "It shall be the duty of owners or occupiers to maintain supply woks in a proper state of repair at their own cost [Para – 25(1)]
   b. "If the Divisional Irrigation Officer is of the opinion that any supply works or part thereof is not properly maintained, he shall, after providing an opportunity to the owner or occupier concerned to carry out maintenance or repair as may be found necessary, carryout such repair or maintenance at the cost of the state government and recover such cost from the owner or occupier, who has failed to carry out the maintenance or repair after due notice, in the manner prescribed. [Para-25(2)]
4. Publication of draft scheme on on-farm development works:
"Not withstanding anything contained to the contrary in this Act and subject to the rules that may be made by the state government in this behalf, an officer designated by the state government may, on his own motion or on the application of not less than fifty percent of the owners or occupiers of lands in the culturable command area, prepare a draft scheme to provide for an on-farm development of a group or block of fields". [Para-27(1)].
5. Declaration to regulate crops:

"Whenever the state government is satisfied that, for better cultivation of lands and due preservation of the water resources of any irrigation work, it is expedient and desirable in the public interest to regulate the kind of crop that should be grown on lands within the culturable command area of any irrigation work and the period of sowing and planting of such kinds of crop, Government having regard to soil characteristics, climate, rainfall and water availability may, by notification, make a declaration to the effect" [Para-35].

6. Assessment of water rate:
"The irrigation officer shall prepare an assessment of the water rates for the lands in respect of which water was supplied, made available or used from an irrigation work, and serve the assessment notice on such owner or occupiers [Para-41].

The Bihar ground water (Regulation and control of development and management) Act, 2006:

Aims & objectives:-

Due to continuous population increase in the country, the exploitation of ground water is also increasing for the purpose of drinking, industrial use and irrigation water. In order to control the exploitation of ground water as per availability, the Government of India had circulated draft model bill for enacting law on control of ground water. The Government of India has been requesting regularly for enactment of such law. This bill was presented with an objective to keep control on exploitation of ground water in those regions of the state where ground water exploitation exceeds its availability. In addition to this, provision is being made to improve ground water status in the critical areas by adopting methods which can conserve the rain water and ensure the recharge of ground water. The control over ground water is essential for meeting needs of future generations.

Salient features of the Act:-
1. "The state government shall, by notification in the official gazette, establish with effect from such date as may be specified in the notification, an authority to be known as the Bihar Ground Water Authority" [Para-3(1)].

2. "The authority shall function under the overall control and supervision of the government" [Para-5(1)]
3. "If the Authority, after consultation with various expert bodies, including Central Ground Water Authority (CGWA) is of the opinion that it is necessary or expedient in the public interest to control and / or regulate the extraction or the use or both of ground water in any form in any area, it will advise the Government to declare any such area to be notified area for the purpose of this Act with effect from such date as may be specified therein. This declaration will be notified in official Gazette. Provided that the date so specified in the notification shall not be earlier than three months from the date of the publication of the said notification" [Para-5(2)]. (Power to notify areas to regulate and control the development and management of ground water).
4. "Any user of ground water as defined under section 2, (1) desiring to sink a well in the notified area of any purpose either on personal or community basis, shall apply to the Authority for grant of a permit for this purpose, and shall not proceed with any activity connected with such sinking unless a permit has been granted by the Authority. Provided that the person or persons will not have to obtain a permit if the well is proposed to be fitted with a hand operated manual pup or water is proposed to be withdrawn by manual devices." [Para-6(1)]. (Grant of permit to extract and use ground water in the notified area).
5. "If the authority is satisfied either on a reference made to it in this behalf or otherwise, that,
    (a) the permit or certificate of registration granted, under sub section (3) of section 6, or sub section (3) of section 7, as the case may be, is not based on facts.
    (b) the holder of the permit or certificate of registration has without reasonable cause failed to comply with the conditions subject to which the permit or certificate of registration has been granted or has contravened any of the provisions of this Act or the rules made there under. Or
    (c) a situation has arisen which warrants limiting of the use or extraction of ground water, then without prejudice to any other

penalty to which the holder of the permit or of the certificate of registration may be liable under this Act, the authority may after giving the holder of the Permit / Certificate of Registration, an opportunity to show cause, cancel the Permit, Certificate of Registration, as the case may be" [(section 10)(a, b, c)] (Cancellation of Permit / Certificate of Registration)

6. Rainwater Harvesting for ground water Recharge:-
"Not withstanding anything contained in the relevant laws, the Municipal Corporation or any other local Authority as the case may be, may impose stipulated conditions for providing roof top rain water harvesting structures in the building plan in an area of 100 sq metres or more while according approval for construction, and permanent water and electricity shall be extended only after compliance of the directions given in this regard."

The Chattishgarh Regulation of Waters Act, 1949:

This Act received the assent of the Governor General on the 10th June 1949 and first published in the "Central Provinces and Berar Gazette, on the 17th June, 1949. The Act was passed to regulate the appropriation of water by industrial concern or a local authority for industrial or urban purposes.

Salient features of this Act:

1. Right of the Government in water:
   "With effect from the date 11 appointed under sub section (3) of section 1 (herein referred to as the 11 relevant date) all rights in the water of any natural source of supply shall vest in the Government except to the extent to which rights may have been acquired in such water before the relevant date, and not withstanding anything contained in the Indian Easement Act 1882(v of 1882) or the Indian Limitation Act, 1908 (ix of 1908) or any custom to the contrary, no rights shall be acquired after such date against the Government by any person in the water from such supply" [Section-3].

2. Restrictions on appropriation of water by an individual concern or legal authority:-

a) "No local authority, which was not appropriating water of any natural source for an industrial or urban purpose immediately before the relevant date, shall, from the said date, appropriate or be entitled to receive for the said propose water from any natural source of supply except in accordance with the rules made in this behalf" [Section 4(1)].

b) "No industrial concern, which has not acquired a right to appropriate water of any natural source for an industrial urban purpose immediately before the relevant date, shall from the said date, appropriate or be entitled to receive, for the said propose water from any natural source of supply except in accordance with the rules made in this behalf "[(Section 4(2)]

3. Private water work:-

"If the state government is satisfied that it is necessary to utilise the water of any private water work, the state government may, by order, direct the owner or other person in charge of such private water work to supply water to a specified industrial concern or local authority in such quantities and in such manner as may be specified in the order" [Section 8(1)].

"The payment of compensation under this section to any person appearing entitled to it in the opinion of the state government or the arbitrator, as the case may be, shall be a full discharge of the state government, the industrial concern or the local authority, as the case may be, from all liability in respect of such compensation but shall not prejudice any rights in respect of the use of the private water-work under this section which any other person may be entitled by due process of law to enforce against the person to whom compensation has been paid as aforesaid". [Sect 8 (5)]

The Karnataka Irrigation Act 1965:

This Act received the assent of the President on 23rd July, 1965. The Act was brought in to make provisions relating to the construction, maintenance and regulation of irrigation works, the supply of water thereon, obtaining labour in emergencies and certain other matters pertaining to irrigation in the State of Karnataka.

Some salient features of this Act are:-
1. Construction, control and maintenance of irrigation works only with consent of Government and subject to conditions:-
"No person shall construct, control or maintain wholly or partly any reservoir, tank, anicut, bandhara, pond, spring pond, canal, field channel, or aqueduct except with previous sanction of the state Government or such other authority as may be authorised by the state Government in this behalf and subject to such conditions as the state Government or such authority may impose.(Provided that the state Government may entrust to a Water Users Society (or Water Users Association) control, maintenance and monitoring of any irrigation work either wholly or in part and there upon such water users society (or water users association) shall be responsible for the control, maintenance and monitoring of such irrigation work)" [Section 4(1)].
2. Notification to issue when water supply is to be applied for purposes of irrigation work:-
"Whenever it appears expedient to the state government or any officer generally or specially authorised by it in this behalf, that the water of any river or stream flowing in a natural channel or of any lake or any other natural collection of still water, should be applied or used by the state government for the purpose of any existing or proposed irrigation work, the state government or the authorised officer may, by notification, declare that the said water will be so applied or used after a day to be specified in the said notification, not being earlier than three months from the date thereof". [Sect 5(1)]
At any time after the day specified under the sub-section (ii), the irrigation officer may enter on any land, remove any obstruction, close any channel and do any other thing necessary for such application or use of the said water, and for such purpose, may take with him or depute or employ such subordinates and other person as he deems fit.
3. Government may prohibit obstruction of river etc within certain limits:-
"Whenever it appears to the state government that injury to the public health, or public convenience or to any irrigation work or to

any land for which supply from an irrigation work is available, has arisen or may arise from the obstruction of any river or stream or natural drainage channel, the state government may, by notification, prohibit within limits to be defined in such notification, or may, within such limits, order the removal or other modification of such obstruction, and there upon so much of the said river, stream or natural drainage channel as is comprised within such limits, shall be deemed to be a drainage work or as defined in section 2" [Section-11].

4. Determination of the need for field-channels and their alignment in any area:-
"The irrigation officer on being satisfied that the construction of field channels in any area is necessary in public interest for supply of water from an irrigation work to lands requiring such supply for purposes of cultivation, shall declare by notification that such field-channels may be constructed after a date to be specified in the notification, not being earlier than thirty days from the date of publication thereof. A copy of such notification shall be sent to the Tahsilder of the area for the publication in the villages concerned" [Sect-15(1)].

5. Deputy Commissioner to acquire land:-
"On receipt of a copy notification water under Section 15, (the Deputy Commissioner shall issue notices to the owner of such land and other persons interested in it to show cause why such land should not be acquired and after giving them a reasonable opportunity of being heard, if satisfied, that such land is required for a field channel, proceed to acquire and take possession of such land under the provisions of the Land Acquisition Act 1894, as if a declaration had been issued by State Government for the acquisition thereof under section 6 of that Act and as if the State Government had thereupon directed the Deputy Commissioner to take order for the acquisition of such land under section 7 of the said Act and as if the state Government had issued orders for immediate possession being taken under section 17 of the said Act" [Section-16(1)].

WATER LEGISLATION IN COUNTRIES

To regulate water and ensure equitable distribution among farming communities settled along the Tiber River in Roman Empire, the first set of water laws came in to existence some 2000 years ago in its primitive form. The Roman water law recognised classes of water rights: private, common & public. The private water was owned by individuals and individual had the right to use it. Common water was the water that everyone has the right to use without limit and without any permission. The public water was the water, which was under state's control. The things like the air, deep sea, running water were termed "res omnium communis". The running water of the stream was not owned by anyone, but when someone took it from steam, it becomes a private property during period of possession.

The common law of England applied the Roman concept that water was owned by all and was 'res communis' and could not be owned by the state or the crown. The laws of 'res omnium communis' gradually got accepted in to law of the Netherlands between 14th & 16th centuries and produced a hybrid law popularly known as Roman-Dutch law.

The arrival of Dutch under the leadership of Jan Van Riebeek in South Africa and settlement in Cape of Good Hope in 1652, involved the application of Roman Dutch law in newly settled community. The evolution of water rights in South Africa was highly influenced by legal developments in the Netherlands.

Since in later of 17th century and whole of 18th century, the Cape of Good Hope was a Dutch colony that was under ultimate control of the Netherlands, the doctrine of state ownership of all public rivers was accepted in the 17th century.

During first years of the Dutch settlement (1652-55), the burghers (settlers) on the up steam of Table Bay Valley, were using the steam water for bathing, washing their belongings. A contingent of the merchants of Dutch East Indian Company fell sick and started having health related problems owing to drinking water sourced from these steams on downstream side. This compelled Van Reebeck to issue a placeact (Place Act) on 10th April, 1655, that prohibited the washing of persons and personal belongings in the stream. The general proclamation on placeact prohibiting up stream pollutions was again repeated in 1657.

Between 1652 and 1740, several placeacts were issued to control the quantitative and qualitative use of the stream water in Table Bay Valley.

On 16th Dec, 1661, Dutch East India Company issued a placeact prohibiting use of water for irrigation by settlers in order to allow company's corn mill to flourish. Later, due to rise in water demand, among settlers in raising their gardens and company millers, led to strife for water, which again forced the company to grant entitlements to regulate water use by turns, so that company milers were not harmed. These entitlements of water use by turns were referred as 'besondere gusti', which allowed upstream use of water for irrigation by farmers by turns on rotation basis.

Water Act No 54 of 1956, Republic of South Africa:
The purpose of this Act was to consolidate and amend laws relating to the control, conservation and use of water for domestic, agricultural, urban and industrial purposes; to make provision for the control in certain respects, of the use of sea water for certain purposes; for the control of certain activities in water in certain areas; for the control of activities which may alter the natural occurrence of certain types of atmosphere precipitation; for the control, in certain respects, of the establishment or the extension of townships in certain areas; and for incidental matters.

The Act defines "Private water" "Public Stream" and "Public Water" as: "Private water" means all water rises or falls naturally on any land or natural drains or is lead on to one or more pieces of land which are the subject of separate original grants, but is not capable of common use for irrigation purposes.

"Public stream" means a national stream of water which flows in a known and defined channel, whether or not such channel is dry during any period of the year and whether or not its conformation has been changed by artificial means, if the water therein is capable of common use for irrigation in two or more pieces of land riparian thereto which are the subject of separate original grants or on one such piece of land and also on crown land which is riparian to such stream. Provided that a stream which fulfils the forgoing conditions in part only of its course shall be deemed to be a public stream as regards that part only;

"Public water" means any water flowing or found in or derived from the bed of a public stream, whether visible or not;

Rights of use of private and public water under this Act are enumerated as under:-

Use of Private water:-

"Subject to the provisions of sub section (2) and of sections 12, 21, 22, 23 and 24 and rights lawfully acquired and existing at the commencement of this Act, the sole and exclusive use and enjoyment of private water shall vest in the owner of the land on which water is found; Provided that nothing in this section contained shall be construed as derogating from the right of an owner of land to a reasonable share of water which, rising on the land of an upper owner, flows in a known and defined channel on, or along the boundary of, land situated beyond that upon which such water rises, and has for a period of not less than 30 years been beneficially used by the owner of the land so situated" [Section 5(1)]

[Sub-s (1) substituted by s.2 of Act no 96 of 1984]

"A person who is, as contemplated in sub section (1), entitled to the use and enjoyment of private water found on any land of which he is the owner, shall not, except under the authority of a permit from the Minister and on such conditions as may be specified in that permit, sell, give or otherwise dispose of such water to any other person for use on any other land, or convey such water for his own use beyond the boundaries of the land on which such water is found"[Section 5(2)]

"The provisions of sub-section (2) shall not apply to-

a) The South African Transport Services and to the selling, giving or disposal in any other manner of any water to the South African Transport services;

b) The conveyance of private water by the owner of the land on which such water is found beyond the boundaries of that land for his domestic purposes, or for the watering of his stock, or in accordance with a direction of the responsible Minister under the Conservation of Agricultural Resources Act, 1983 (Act no 43 of 1983), or in implementing any written advice from a competent officer of the state regarding the controlled run off or drainage of rain water for purposes of soil conservation or improvement;

(c) any owner of land situated in the area of any municipal institution contemplated in section 84 (1) (f) of the Provincial Government Act, 1961, or a local authority established under section 2 of the Black Local Authorities Act, 1982" [Section -5 (3)]
[Sub –s. (3) substituted by s. 2 of Act No. 92 of 1980 and by S.4 (a) of Act No 68 of 1987]
Use of public water by authorized persons for certain purposes:

Notwithstanding anything contained in this Act:
a) Any person may, while he is lawfully at any place where he has access to a public stream, take and use water from such stream for the immediate purpose of watering stock or drinking washing or cooking, or use in a vehicle at that place;
b) Any officer or servant of a provincial administration, divisional council or other lawfully constituted body which is responsible for the construction, maintenance, repair or control of any road (other than a road within the area of jurisdiction of any municipal or other like institution) may, while he has access to a public stream, take and use so much of the water from such stream as may be necessary for the purpose of constructing, maintaining, repairing or controlling such road, provided no riparian owner along the course of such stream is thereby deprived of water from such stream for his own use for the irrigation of land which is then under irrigation or for domestic purposes or for the watering of his stock. [Section -7]

Use normal flow a public stream:
"Subject to the provisions of this Act, and to any existing right, every riparian owner is entitled to the reasonable use of such, as may have been lawfully acquired by him, from any other person and of his share (as determined under section fifty-two) of the normal flow of a public stream, to which his land is riparian for use for agricultural and urban purpose on such land; Provided that-
a) A riparian owner shall not use such water wastefully or detain any portion thereof unreasonably or allow it to runoff in unreasonable quantities to the detriment of lower owners;

b) A riparian owner who uses any portion of such normal flow for agricultural purposes shall on his own land, if practicable, or otherwise at the nearest convenient point elsewhere, return such water to the public stream from which it was abstracted with no other loss than that which has been occasioned by such use, unless he is legally obliged to pass the water on to other land;

c) A riparian owner shall not use any portion of such normal flow for the irrigation of land, if thereby he deprives any lower riparian owner having a right to the use of such normal flow, other than an inhabitant within the area of jurisdiction of a local authority who is being supplied with water by that local authority, of water for domestic purpose or the watering of stock;

d) A lower riparian owner referred to in paragraph (c) shall satisfy his needs as to domestic use and the watering of his stock out of so much of the normal flow as he is entitled to use for agricultural purposes before he shall be entitled to demand, for the purpose of his domestic use or the watering of his stock, that any other riparian owner shall abate any of the share of the normal flow such last mentioned owner is entitled to use for agricultural purposes;

Subject to the provisions of this Act and to any existing right, the owners of land riparian to a tributary to a public stream shall be entitled to so much of the normal flow of such tributary as they may require for any of the aforesaid uses in preference to any right of any of other owner to the normal flow of such tributary for such uses.

Water laws in Mexico:-
During the colonial period, the subject of water was the domain of Spain Crown which permitted granting of concessions or permits for utilization of water. In 1957, the Mexican constitution was empowered to determine which waters were under federal jurisdiction and was empowered to enact laws relating to water uses. However, fifth paragraph of Mexico's 1917 Constitution defines which are "national waters" under Federal Government's jurisdiction and read as:

"The nation has domain over waters in territorial seas in the extent and terms provided in international laws; interior marine waters, lagoons and estuaries permanently or intermittently connected with the sea; naturally formed in land lakes directly connected with constant flow currents; rivers and their direct and indirect tributaries, as of the place in the river bed where permanent, intermittent or torrential flow of water starts and up to river's mouth in the national property seas, lakes, lagoons or estuaries; Constant or intermittent water flows and their direct or indirect tributaries, when the river bed throughout its length or part thereof is a country border or is a boundary between two states or when it crosses state lines or country's borders; lakes lagoons and estuaries whose basins, zones or shores are crossed by the borders to two or more states as the republic and a neighbouring nation; or when the shore line is the border of two states of the Republic and a neighbouring nation; springs in beaches, maritime zones, river beds, riverbanks or lakeshores and inland currents to the extent determined by the law, underground waters may be freely drilled through manmade works and appropriated by land owner, but when public interest requires or other users may be affected, the federal executive may regulate their extraction and use and even specify "off limits" zones, as well as for any other national property waters. Other waters not included in the forgoing list shall be considered integral parts of the land property in which they flow or are deposited but if are located in two or more parcels of land, they shall be deemed of public interest and shall be subject to regulations enacted by the states".

Therefore, it is evident that, practically all water in Mexico is deemed "National Water", except the rainfall before it falls on the ground.

The sixth paragraph further states:- "In the cases referred to in the two previous paragraphs (mineral resources and national waters, respectively, the Nation's domain is an inaliable right not subject to adverse possession and the exploitation, use or utilization of the resources mentioned therein by private individuals or corporate entities incorporated under the Mexican laws, can only be done through concessions granted by the Federal Executive in accordance with the rules and conditions established in the laws".

Irrigation Law of 1926 was the first law application to "national waters". This law was superseded by "Federal Waters Laws" of 1972. The existing "National Waters Law" (NWL) was published in official Gazette of Mexico (Diario oficial de la Federation) on December 1, 1992 and became effective from December 2, 1992.

It is considered the most important piece of water legislation. It contains 124 articles in 10 titles/ Sections, namely: (1) General Provisions (2) Water Management (3) Hydraulic Programming (4) Rights to use or utilization of National waters (5) Regulated off limits and Reserve zones (6) Uses of water, Infrastructure (9) National Property entrusted to the National Water Commission and (10) Violations, Penalties and Recourses.

The main purpose is to regulate the exploitation and use or utilization of national waters, its distribution, and control, quantitative and qualitative conservation to achieve sustainable development. It confirms the national control over national waters. It also provides that all users of national waters are obliged to pay fees for use of national waters.

National Irrigation Policy of Brazil:
The national Irrigation Policy of Brazil was implemented through Law no. 12,787 of 11th January, 2013.The law no. 12,787 defines various categories, such as, "farmer irrigator", "family-based irrigator farmer," "irrigated agriculture" etc. The irrigation policy is founded on following principles:
i) Sustainable use and management of land and water resources for irrigation.
ii) Integration with specific policies on water, environment, energy, environmental sanitation, rural credit and insurance and their respective plans, with priority given to projects whose works allow multiple uses of water resources
iii) The linking of the actions concerning irrigation in different instances and levels of government and between them and the actions of the private sector.
iv) Democratic and participatory management of public irrigation projects with irrigation infrastructure in common use through mechanisms to be defined by regulation.
v) Prevention of water borne rural endemics.

The main objectives of the irrigation policy are:
1) To encourage the expansion of irrigated areas and increase productivity in an environmentally sustainable basis.

2) To reduce climate risks inherent in agricultural activities particularly in areas of low rainfall.
3) To promote local and regional development.
4) To train human resources and foster the creation and transfer of technologies related to irrigation.
5) To encourage private irrigation projects as defined by regulation.

The water rights are enumerated in the Water Code under various articles stated under "codigo de Agus, Decreto No 24643 10 de Julho de 1934" (Decree No 24,643 of July 10, 1934) where in following articles give stated rights as referred.

Article 89 "For purpose of the Water Code (Eodigo de Aguas), a spring (nascente) is defined as water that emerge naturally or by human industry, and run within one particular property, and even go through it, when those waters have not been abandoned by the owner"

Article 90 "The owner of the property where spring is located cannot impede the natural course of water to lower properties once the owner's water consumption needs have been satisfied"

Article 91 "If spring emerges in a gap that divides two properties, it belongs to both properties"

Article 92 "According to the norms of legal drainage servitude (normas da servidao legal de escoamento), the owners of low-lying properties are compelled to receive water from artificial sources (nascentes artificials) provided that they receive compensation.

("This compensation, however, must take in to action the value of any benefit that can be derived from supplying such water to the properties")

Article 94: "The owner of a spring cannot alter its course when the spring supplies water to a population"

Article 95: "Spring water is determined by the point at which it begins to run upon the ground and not by the underground vein that feeds it".

Following articles relate to ground water:
Article 96: "The owner of any land will be able to appropriate through wells, galleries etc, of waters that exist under the surface of his building,

as long as it does not harm existing users or derives or diverts from his natural course public water, public in common use or private"

("If the use of groundwater referred to in this article damages or reduces public or public water in common or private use, the competent administration may suspend said works and use")

Article 97: The owner of the building cannot open a well next to the neighbour's building, without keeping the necessary distances or taking the necessary precautions so that he does not suffer any damage"

Article 98: Buildings that are capable of polluting or rendering useless for the ordinary use, the water from the well or spring that belongs to them, or pre- existing.

Article 99: Anyone who violates the provisions of the previous article is obliged to demolish the constructions made, responding for losses and damages.

Article 100: "The currents that momentarily disappear from the ground, forming an underground course, to reappear further away, do not lose the character of a public thing in common use, when they were already in its origin".

Article 101: "The openings of wells on land in the public domain depend on an administrative concession.

Articles related to Rain water:
Article 102: "Rain water is considered to be that which comes immediately from the rains.

Article 103: The rain water belongs to the owner of the building where it falls directly, and he can dispose it at will, unless there is a right to the contrary.

The building owner however, is not allowed to; 1st, waste these waters to the detriment of other buildings that can take advantage of them, under penalty of compensation to their owners; 2nd, divert these waters from their natural course to give them another, without the express consent of the owners of the buildings that will receive them.

Article 104: "Overcoming the boundary of the building in which rainwater falls, abandoned by its owner, as applicable to them, the rules dictated for common waters and public are subject".

BIBLIOGRAPHY

1. "Monsoon Prediction" - Dr. Ajit Tyagi, D.R.Pattanaik, "Yojana", July'2012.
2. "South West Monsoon in India and its Forecasting System" – S.C.Bhan, "Yojana", July'2012.
3. Irrigation and Water Power Engineering By Dr. BC Punmia& Dr. Pande B.B. Lal, Standard Publishers Distributors,New Delhi.
4. http://simple.wikipedia.org/wiki/rain
5. Monsoon – By Dr. Subrata Bose and Dr. Neeraj Saxena, Employment News weekly, 26may-1June-2001.
6. "The next major conflict in the middle east? Water wars" – Adel Darwish, Geneva conference on environment and quality of life – June'1994.
7. "Water resources-future scenario, issues and options" – Jay Narayan Vyas, Shri V.Raman Endowment Lecture, 36th National Annual Convention of Indian Water Works Association (IWWA), 9th Jan'2004, Ahmedabad.
8. Water harvesting structures planning, design and construction, Chapter - VI – www.mphed.org.
9. "Water Conflict Chronology" – Dr. Peter H. Gleick. Pacific Institute for studies in Development, Environment and security.
10. Kurukshetra, February'2011.
11. Shereen Jegtvig – About.Com guide.
12. http://www.fao.org/nr/aquastat
13. http://www.cwc.nic.in/main/webpages/statistics.html
14. "Virtual water trade and geopolitics", World Water Forum, Kyoto, 2003, Daniel Zimmer, Director-World Water Council.
15. "Status of Virtual water trade from India", Current Science, vol-93, Oct'2007. "Climate change and Global Water Crisis: What Business Need to Know and Do", May'2009, Pacific Institute-UN Global Compact.
16. UN World Water Development Report 4(WWD4) -2012.

17. "Environmental Protection: Important Tips for the Development of Agroforestry" – by R.S.Sengar and Reshu Chaudhary, Kurukshetra-A Journal on Rural Development. vol- 60, June 2012.
18. "Environment and Economic Development: A correlation", byBarnaMaulick - Kurukshetra-A Journal on Rural Development. vol- 60, June 2012.
19. "China, Tibet and the Strategic Power of Water: Pollution and Global Warming Threaten Asia's Most Important Fresh Water" – Circle of Blue, http/www.csrwire.com/press_release/, dtd-May-08, 2008.
20. JEEVA et al: Traditional Agricultural Practices of Meghalaya, Indian J Traditional Knowledge, Vol.5, No.1, January 2006.
21. http://www.unep.org/dewa/vitalwater/article115.html
22. "Nitrogen Fixing Trees: Multipurpose Pioneers" – Craig Elevitch and Kim Wilkinson, Permaculture International Journal, Issue no-56.
23. "Water Resources Management in 21st Century" – Dr. TayebAbdullabhaiSihorwala, Journal of IWWA, July-Sept' 2000.
23. "WHO Guidelines for the Safe Use of Waste Water for Irrigation" – Dr. P.G.Sastry, Journal of IWWA, January – March'2009.
24. "Socio-economic aspects of water" – Prof. P.B.Sharma, Journal of IWWA, July –Sept.'2000.
25. "A case study of Waste Water Recycling Plant at South Western Railways, Bangalore" – K.R.Sree Harsha, Dr. L.U.Simha, C.T.Puttaswamy, - Journal of IWWA, Oct-Dec'2009.
26. "Water Conservation by control of Evaporation and Seepage" by Shri V.B.Patel and Shri D.T.Buch – Best of IWWA (Selected compilation of Journal of IWWA articles),1994.
27. "Evaporation Control in Reservoirs", Central Water Commission, Basin Planning and Monitoring Organisation, New Delhi, 2006.
28. ttp://www.globalwaterintel.com/archive/7/12/market-insight/turning-losses-into-gains.html.
http://www.switchurbanwater.eu/outputs/pdfs/GEN_PRS_Leakage_Management_and_Control_AC_Apr08.pdf
29. "Manila Water NRW Reduction story"   Abelardo P. Basilio, IWWA Journal,July-December'2008.
30. "Managing Water Demand in Singapore",Ramahad Singh, P.U.B, Singapore, in Journal of IWWA, July-December'2008.
31. "Reduction of Apparent Losses Using the UFR (Unmeasured Flow Reducer) – Case Studies. Sharon Yaniv, Journal of IWWA, July-December'2008.

32. "Benchmarking of Losses from Potable Water Reticulation Systems- Results from IWA Task Team"- R.SMckenzie, C.Seago and R.Liemberger, Journal of IWWA, July-December'2008.
33. "Water loss levels from Transmission Mains in Urban Environment", Cliff Jones and Kevin Laven, Journal of IWWA, July-December'2008.
34. www.puretechltd.com/products/sahara/sahara_video.shtml
35. "Consequences of Rural Migration", Dr. Parveen Kumar, Kurukshetra – Journal on rural development, September'2014.
36. "Environmental Refugees- The result of another form of forced rural migration", Dr. Srikanta K. Panigrahi, Kurukshetra – Journal on rural development, September'2014.
37. "Virtual Water in Food Production and Global Trade Review of Methodological Issues and Preliminary Results", Daniel Jimmer and Daniel Renault, World Water Council, FAO-AGLW.
38. "Virtual Water: An Introduction", A.Y.Hoekstra, 'Value of Water Research Report Series No.12, February'2003, IHE, DELFT, Netherlands.
39. "Virtual Water Trade: A qualification of Virtual Water flows between nations to international crop trade", A.Y.Hoekstra and P.Q.Hung, 'Value of Water Research Report Series No.12, February'2003, IHE, DELFT, Netherlands.
40. "The role of water harvesting in alleviating water scarcity in arid regions", Prof. Dr. Dieter Prinz, Institute of Water Resources Management, University of Karlsruhe, Germany.
41. "How Cape Town Went from Water Crisis to Overflowing Dams in Just 2 Years", Khanyi Mlaba, "Global Citizen", Oct 09, 2020.
42. "Mexico City Keeps Sinking As Its Water Supply Wastes Away", Carrie Kahn, 14th September'2018, npr.
43. "National Compilation on Dynamic Ground Water Resources of India, 2017", Ministry of Jal Shakti, July'2019.
44. "Water scarcity in Beijing and countermeasures to solve the problem at river basins scale", Lixia Wang et al, Nanjing Institute of Environmental Sciences, Nanjing, China.

HORIZON

The Assam Tribune, GUWAHATI

FRIDAY, MARCH 29, 2019

FEELINGS 2 — Unexpected twist
ODD WORLD 3 — Living in the present

Leaving no one behind

Seven hundred million people worldwide could be displaced by intense water scarcity by 2030, writes MRIDUL DEKA.

More women, more equality

Pallavi Goswami

"Leaving no one behind"
Mridul Deka
(As published in "The Assam Tribune",
Guwahati, dtd-29-03-2019)

A seven-year-old child Jakelin Caal from Guatemala died in custody of U.S Border Patrol when her father took her along to cross over to U.S side of the Border in December'2018. A week later her mother Claudia Macquin told Jeff Abbot and Sandra Cuffe, two reporters from Aljazeera, that the family survived on sustenance agriculture, farming corn and other staples in a small patch of land they owned and that their family made about $90 over a period of six months from selling corn harvest, left over after feeding the family. Her husband and her daughter Jakelin left home to find a better life in the United States. "All lands are in the hands of large land owners and palm companies", fumed Jakelin's mother in deep anguish. Between 2001 and 2012, there has been 207 percent increase in land use for palm plantations in Guatemala. Palm has become a lucrative business for many companies over the years and 18 major rivers and tributaries that flow through Guatemala in to Pacific Ocean, many of these water ways have been diverted to run through palm and sugarcane plantations leaving many Q'eqchi communities, to which Jakelin belong, throughout eastern and northern Guatemala high and dry with scant land ownerships and concomitantly failing water sources.

The Ujjani Dam in Solapur District of Maharashtra in India has been providing water for the farmers in the region since its inception in 1980. However, of its 110 TMC (Thousand Million Cubic Feet) of water, 60 TMC is being illegally diverted to sugarcane fields, creating water shortage in hundreds of villages in eight taluks of Madha, Pandharpur, Mohol, Mangalvedha, Malshiras, North Solapur, South Solapur, Akkalkot of Solapur district of the state. Whereas Ujjani's water supply is reserved for crops like chilli, jower, bajra, ground nut, maize, sun flower, wheat, gram, and vegetables, 51 percent of its supply goes to six lakh hectares of sugarcane fields spread across three districts. These three districts have as many as fifty sugarcane factories, most of which are owned by politicians or cooperative societies chaired by politicians having allegiance to National Congress Party (NCP), wrote Kiran Tare, Senior Associate Editor, "India Today" in July'2012. Even though the Govt. records state that only 4 percent of the land in Maharashtra is under sugar cultivations, what is not pronounced is 71.5 percent of all irrigated water reaches sugar plantations. South Asia Network on Dams, River & People (SANDRP) says if 50 percent of the water being used to cultivate sugarcane in the region was diverted for production of pulses, it would mean livelihood security to over 21 lakhs farmers as against 1.1 lakhs sugarcane farmers supported now.

These two are among numerous examples of how indigenous communities are being systematically deprived of water resources by more elite and powerful section of the society they belong. Anne Platt of World Watch Institute reports that a family in top fifth income groups in Peru is three times more likely to have a piped water supply connection in their homes than a family in bottom fifth income group. Similarly, a family in top fifth income group in Dominican Republic or Ghana is six and twelve times more likely to have a piped water connection in their premises than a family in bottom fifth income groups in those countries respectively. In Dhaka, Bangladesh, squatters pay water rates as much as 12 times higher than what the local utility charges. A low-income family in Lusaka, Zambia spends half of their income on water. Elsewhere in East, when a major drought hit Indonesia in 1994, while people in the country experienced severe water shortage due to their wells running dry, Jakarta's Golf courses catering to wealthy tourists continued to receive 1 million litres of water per course per day. And just when most of the rivers dried up resulting in depletion of aquifers throughout the land causing acute scarcity of water resources, the Govt. of Cyprus cut the water supply to farmers by 50 percent in 1998 while guaranteeing the country's two million tourists a year all the water they needed. Where race and class matter, water privilege can be shocking. In South Africa, six lacs white farmers consume nearly 60 percent of the country's water supplies for irrigation, while 15 million blacks have no direct access to water, writes Maude Barlow of Council of Canadians, Canada's largest public advocacy organization. And in Plachimada, Kerala in India, it took Supreme Court to intervene in 2005, in a water dispute between Virudha Janakeeya Samara Samity, a representation of local adivasis and the Hindustan Coca Cola Beverages Pvt. Ltd and reduce the ground water withdrawal by one-third of its earlier permitted limit of 1.5 million litres to produce 14,400 bottles of 300ml soft drink daily, after the local tribal communities complained of their wells going dry and the wastes from the factory damaging their crops due to excessive withdrawal of water by the company.

A fact sheet released by United Nations to commemorate the World Water Day 2019, states, globally, 2.1 billion people live without safe water at home. More than 700 children under five years of age die every day from diarrhoea linked to unsafe water and poor sanitation. The fact sheet also states that, one in four primary schools have no drinking water service, with pupils using unprotected sources or going thirsty. Eighty percent of the people who have to use unsafe and unprotected water sources live in rural areas. Around 4 billion people nearly two third of the world's population- experience severe water scarcity for at least one month of the year. And Seven hundred million people worldwide could be displaced by intense water scarcity by 2030.

As the world celebrated "World Water Day" this year again on 22nd March, The UN had declared 'Leaving no one behind' as the theme for 2019, which is an expression of Sustainable Development Goal 6 (SDG6), aiming to ensure availability and sustainable management of water for all by 2030. This means *leaving no one behind* including those marginalized groups such as women, children, refugees, indigenous people, disabled and many others as mentioned who are often overlooked and face discrimination as they try to access and manage their safe water needs.

(Written on** the **occasion of 'World Water Day'-22nd March'2019)

The Sentinel

of this land, for its people

TUESDAY 9 AUGUST 2016

Doping slurs mar Rio Games

The medal rush in Rio Olympics is picking up momentum, with the most decorated Olympian Michael Phelps already getting into the action with his 19th gold. But the fallout continues over doping slurs, even becoming a part of mind games some athletes are employing in the Games. The swimming pool is abuzz with a spat between the Australian and Chinese swimming federations, after Mack Horton beat defending champion Sun Yang in the 400m freestyle, and then called Yang a drug cheat. The reference was to a secret three-month ban Yang served in 2014 after he tested positive, but the Chinese establishment had then kept the entire matter under wraps. India too has got its share of jolts when shotputter Inderjeet Singh tested positive and was barred, while grappler Narsingh Yadav booked a last gasp ticket to Rio after it was adjudged his food had been spiked with drugs by a rival. The case of the Russian contingent is far more serious, with over 100 athletes barred from the Games due to doping allegations. Its entire track and field team was served an unprecedented ban by the IAAF, after the scandal broke just a fortnight before the Games of a massive state-sponsored doping and cover-up in Russia. The World Anti-Doping Agency (WADA) report prompted Olympic associations of the US, Canada and several other countries to demand that the entire Russian contingent should be thrown out from the Games.

The International Olympics Committee decided against a blanket ban, instead putting the onus on the 28 individual sports federations that oversee the Olympic disciplines, to furnish lists of 'approved' athletes. This smacked of an attempt to pass the buck by the IOC, which in turn has exposed deep fissures between and within top sporting bodies. IOC chief Thomas Bach was accused of being chummy with Russian president Vladimir Putin, who has been running a strident campaign that athletes from his country are being systematically and unfairly targeted. The IOC chief has castigated WADA for releasing the report so close to the Games and creating chaos thereby, calling for a full review of its anti-doping system. But WADA has claimed the report was published only after it was in their hands.

Water uncertainties in riparian countries

By Mridul Deka

Two days after the Egyptian intelligence leaked an information to press in the month of November 1989 about Israeli hydrologists studying some areas on Blue Nile for construction of few dams on Ethiopian soil to store 51BCM (Billion Cubic Meters) of Nile waters, the Ethiopian ambassador was called to the foreign office in Cairo to provide an explanation on the matter. Understandably, Egyptian side cautioned the Ethiopian ambassador against any interference of Egypt's inherent rights to Nile waters through a Nile River Agreement which Great Britain had signed with Egypt in 1929. The Anglo-Egyptian Treaty granted Egypt an annual water allocation of 48 BCM and Sudan, a share of 4 BCM out of the estimated average annual yield of Nile River at 84 BCM. The 1929 agreement also gave the Egyptian authorities, the veto power over all construction projects on river Nile or its tributaries. Through another treaty signed between Egypt and Sudan in 1959, the water allocation of river Nile was raised from 48 BCM to 55.5 BCM for Egypt and for Sudan, water allocation was increased from 4 BCM to 18.5 BCM. What both parties conveniently forgot was, there were at least nine other riparian countries which also depend on Nile waters for their survival. Quite plausible is the fact that upper riparian states like Kenya, Tanzania, Uganda and Ethiopia do not recognize these agreements any longer. In fact, Tanganyika (now Tanzania) was the first country to oppose these agreements raising its concern shortly after it gained independence from Great Britain in 1961.

River Nile itself has two tributaries White and the Blue Nile which meet at Khartoum in Sudan. While the White Nile originates in Lake Victoria, Blue Nile originates in Lake Tana in Ethiopian highland. It is the Blue Nile which contributes more than 80 percent of discharge to the river Nile. Therefore Ethiopia's demands and concerns for an equitable share of Nile waters are seemingly reasonable. Ethiopia has about 3.7 mha (million hectares) of land, which can be irrigated. It is estimated that irrigating fifty percent of this land will reduce river Nile's flow by 15 percent. For centuries, River Nile has been a cultural symbol of Egypt. In fact, the river is the only source of water for 40 million farmers of Egypt irrigating their fields in this desert nation. Therefore, it is obvious, that any diversion of Nile waters would seriously harm Egypt, which the country will not tolerate. Until recently, Egypt had little or no worries for its share of fresh water from the Nile. However, Egypt's complacencies were lost when erstwhile Prime Minister of Ethiopia Meles Zenawi, a member of Chinese Academy of Sciences, Wang on 2 April, 2011. Formerly called the Millennium Dam, the Grand Ethiopian Renaissance Dam on Blue Nile is located at about 15 Kms east of Sudanese border. Presently, half way through, this gravity dam when complete, will have storage capacity of 79 BCM and would house the largest hydroelectric power plant in Africa and would be generating 6000 MW of electricity. This project with initial cost of US$4.8 Billion is slated for completion in July 2017.

Yellow River, the longest in China traverses through Tibetan Plateau (Upper basin), the Loess Plateau (Middle basin) and the North China Plain (Lower basin). Nearly 70 percent of the basin population resides in the lower third of the basin. Many industrial cities like Taiyang,

Yinchuan, Baotou, Zhengzhou, and Jinan came up in the basin giving rise to tremendous water demand. The river carries 1.3 billion tons of sediments each year. The salt deposits in the river bed and consequent rise in its levels has made approximately 90 million people in the basin vulnerable to floods. Tibet acts as the fresh water source of approximately 85 percent of Asian population. It is estimated that Tibetan plateau stores up to 12,000 cubic kilometers of fresh water. Its fresh water reserves are so bountiful that it serves as the head waters of many of the Asia's largest rivers including Yellow, Yangtze, Mekong, Brahmaputra, are among others. Almost half the world population lives in the water sheds of these rivers. However, Intergovernmental Panel on Climate Change (IPCC)'s report in May 2007 highlights the fact that Tibet's glaciers are receding at a rate of 0.9 meters annually, and at least 500 million people in Asia and 250 million people in China are at risk from declining glacial flows on Tibetan plateau.

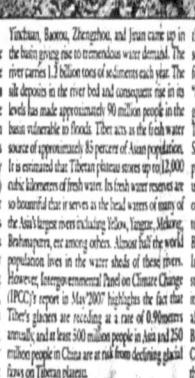

diverts water from the upper reaches of the Brahmaputra River in Tibet to country's northern province of Xinjiang. The newly proposed route is expected start from the Brahmaputra River and would carry water to Xinjiang along the Qinghai-Tibet Railway and Hexi Corridor. However, the new plan proposed by Wang Guangqian is inspired by the "Shouwan Canal" conceived by Chinese hydro-geologist Gao Kai in 1988. Also called the Tianguo Project, the Shouwan Canal aims to divert 200 billion cubic meters comprising nearly 33 percent of Brahmaputra River's water to Yellow River each year. The proposed dam of the project at Shoumatan Point on Great Bend would have the world's largest hydroelectric facility. It is believed that there will be much more diversion than the originally planned 200 BCM, as the recent studies point to faster glacial melt than earlier projections of 1990s when climate change considerations were not applied. Despite repeated denial by Chinese authorities, reports and opinions suggest that the Yarlung Tsangpo diversion project had already been initiated with allocation of 750M Yuan in its 11th 5 year plan out of the 100B Yuan for capital works projects in Tibet for construction of the Medog highway. Roughly 20 kilometers to north of Medog is the site for the proposed storage dam of the Tsangpo Project.

Like river Nile and Brahmaputra and in similar cases around the world involving riparian states, the upper riparian state will always have a strategic advantage and may be tempted to use it as a political tool to put pressure on a lower riparian nation. Not only a diversion dam on the upper reaches can create scarcity in a downstream state, but it can also cause flood havoc through sudden release of excess water during rainy seasons. Not to forget, in 2000, General Zhao Nanqi, former Director of People's Liberation Army, declared "even if we do not build this water diversion project, next generation will, sooner or later, it would be done". Therefore, it would be prudent on the part of our Civil Societies and NGOs to voice their concerns about the proposed dam/plans on Yarlung Tsangpo or Brahmaputra under Chinese control than protesting against dams/diversion over Brahmaputra in India. As what river Nile means to people of Egypt, similar is the importance of river Brahmaputra to the land and culture of people of Assam in India. With a stable Government at centre and a rising stature in world affairs, let us hope that the nation will use its influence and persuade China to refrain from going ahead with any water diversion projects on river Brahmaputra, be it Tsangpo Project or the Great Western Canal project so that unlike the Egyptians, present and the future generations of Assam, do not have to live with water uncertainties. *(Mridul Deka is former Advisor (Engg), CWC)*

Water Uncertainties in Lower Riparian Countries.
Mridul Deka
(As published in "The Sentinel", Guwahati, dtd- 09-08-2016)

Two days after the Egyptian intelligence leaked an information to press in the month of November'1989 about Israeli hydrologists studying some areas on Blue Nile for construction of few dams on Ethiopian soil to store 51BCM (Billion Cubic Meters) of Nile waters, the Ethiopian ambassador was called to the foreign office in Cairo to provide an explanation on the matter. Understandably, Egyptian side cautioned the Ethiopian ambassador against any interference of Egypt's inherent right to Nile waters through a Nile River Agreement which Great Britain had signed with Egypt in 1929. The Anglo Egyptian Treaty granted Egypt an annual water allocation of 48 BCM and Sudan, a share of 4 BCM out of the estimated average annual yield of Nile River at 84 BCM. The 1929 agreement also gave the Egyptian authorities, the veto power over all construction projects on river Nile or its tributaries. Through another treaty signed between Egypt and Sudan in 1959, the water allocation of river Nile was raised from 48 BCM to 55.5 BCM for Egypt and for Sudan, water allocation was increased from 4 BCM to 18.5BCM. What both these parties conveniently forgot was, there were at least nine other riparian countries which also depend on Nile waters for their survival. Quite plausible is the fact that upper riparian states like Kenya, Tanzania, Uganda and Ethiopia do not recognize these agreements any longer. In fact, Tanganyika (now Tanzania) was the first country to oppose these agreements raising its concern shortly after it gained independence from Great Britain in 1961.

River Nile itself has two tributaries White and the Blue Nile which meet at Khartoum in Sudan. While the White Nile originates in Lake Victoria, Blue Nile originates in Lake Tana in Ethiopian highland. It is the Blue Nile which contributes more than 80 percent of discharge in the river Nile. Therefore, Ethiopia's demands and concerns for an equitable share of Nile waters are seemingly reasonable. Ethiopia has about 3.7 mha (million hectors) of land, which can be irrigated. It is estimated that irrigating fifty percent of this land will reduce river Nile's flow by 15 percent. For centuries, River Nile has been a cultural symbol of Egypt. In fact, the river is the only source of water for 40 million farmers of Egypt irrigating their fields in this desert nation. Therefore, it is obvious that any diversion of Nile waters would seriously harm Egypt, which the country will not tolerate. Until recently, Egypt had little or no worries for its share of fresh water from the Nile. However, Egypt's complacencies were lost when erstwhile Prime Minister of Ethiopia Meles Zenawi laid the foundation of Grand Ethiopian Renaissance Dam on 2nd April, 2011. Formerly called the Millennium Dam, the Grand Ethiopian Renaissance Dam on Blue Nile is located at about 15Kms east of Sudanese border. Presently half way through, this gravity dam when complete, will have storage capacity of 79 BCM and would house the largest hydroelectric power plant in Africa and would be generating 6000 MW of electricity. This project with initial cost of US $ 4.8 billion is slated for completion in July 2017.

Yellow River, the longest in China traverses through Tibetan Plateau (Upper basin), the Loess Plateau (Middle basin) and the North China Plains (Lower basin). Nearly 70 percent of the basin population resides in the lower third of the basin. Many industrial cities like Xining, Yinchuan, Baotou, Zhengzhou, and Jinan came up in the basin giving rise to tremendous water demand. The river carries 1.2 billion tons of sediments each year. The silt deposits in the river bed and consequent rise in its levels has made approximately 90 million people in the basin vulnerable to floods. Tibet acts as the fresh water source of approximately 85 percent of Asian population. It is estimated that Tibetan plateau stores up to 12,000 cubic kilometers of fresh water. Its fresh water reserves are so bountiful that it serves as the head waters of many of the Asia's largest rivers including Yellow, Yangtze, Mekong, Brahmaputra, etc among others. Almost half the world population lives in the water sheds of these rivers. However, Intergovernmental Panel on Climate Change (IPCC)'s report in May'2007 highlights the fact that Tibet's glaciers are receding at a rate of 0.90meters annually, and at least 500 million people in Asia and 250 million people in China are at risk from declining glacial flows on Tibetan plateau.

A member at Chinese Academy of Sciences, Wang Guangqian in June'2011, raised a new proposal to divert water from the upper reaches of the Brahmaputra River in Tibet to country's northern province of Xinjiang. The newly proposed route is expected start from the Brahmaputra River and would carry water to Xinjiang along the Qinghai-Tibet Railway and Hexi Corridor. However, the new plan proposed by Wang Guangqian is inspired by the "Shoutian Canal"- conceived by Chinese hydro-geologist Guo Kai in 1988. Also called the Tsangpo Project, the Shoutian Canal aims to divert 200 billion cubic meters comprising nearly 33 percent of Brahmaputra River's water to Yellow River each year. The proposed dam of the project at Shoumatan Point on Great Bend would have the world's largest hydroelectric facility. It is believed that there will be much more diversion than the originally planned 200 BCM, as the recent studies point to faster glacial melt than earlier projections of 1990 when climate change considerations were not applied. Despite repeated denial by Chinese authorities, reports and opinions suggest that the Yarlung Tsangpo diversion project had already been initiated with allocation of 750M Yuan in its 11th 5-year plan out of the 100B Yuan for capital works projects in Tibet for construction of the Medog highway. Roughly 20 kilometers to north of Medog is the site for the proposed storage dam of Tsangpo Project.

Like river Nile and Brahmaputra and in similar cases around the world involving riparian states, the upper riparian state will always have a strategic advantage and may be tempted to use water as a political tool to put pressure on a lower riparian nation. Not to forget, in 2000, General Zhao Nanqi, former Director of People's Liberation Army, declared "even if we do not build this water diversion project, next generation will… sooner or later it would be done". Therefore, it would be prudent on the part of our Civil Societies and NGOs to voice their concerns about the proposed dam/dams on Yarlung Tsangpo or Brahmaputra under Chinese control than protesting against dam/dams on river Brahmaputra in India. Because what river Nile means to people of Egypt, similar is the importance of river Brahmaputra to life and culture of people of Assam in India. With a stable Government at centre and a rising status in world affairs, let us hope that the nation will use its influence and persuade China to refrain from going ahead with any water diversion projects on river Brahmaputra, be it Tsangpo Project or the Great Western Canal project so that unlike the Egyptians, we and our future generations do not have to live with water uncertainties.

horizon

Saturday Special

The Assam Tribune, Guwahati, Saturday, December 17, 2005

Although water is found abundantly on planet Earth, a majority of its population do not have access to clean drinking water. Guwahati too, despite being situated on the banks of the mighty Brahmaputra, is no exception with a regular shortfall of water, writes **Mridul Deka.**

Water WOES

Six out of ten bore wells dug in north Gujarat and Saurashtra Kutch region yield no water even at a depth of 1200 ft. In Loej village of Gujarat, each of the 200 families has one member suffering from kidney problem. In Rajasthan, ten towns get water once in three days, and 31 once in two days. Twelve towns get it from 200 km away. Sixty-six million people in 200 districts risk fluoride contamination and 15 million risk arsenic poisoning. (India Today: June 9, 2003)

Can we be an economic powerhouse without water? That's the question the country should be asking after per capita availability of fresh water has fallen from 5177 cubic metres to 1869 cubic metres in 50 years. That's perilously close to 1700 mark fixed by the United Nations, below which we will become a 'water-stressed' nation. (Business Today, June 6, 2003)

Alarming statistics? Add some more to the list:

■ About 1.1 billion people across the globe lack access to clean drinking water.

■ More than 200 towns in India, excluding the classI cities, get less than 70 lpcd (litres per capita per day) of clean water. Compare this to 262 lpcd, average water consumption by US citizens.

■ Union Ministry of Water Resources estimates that water demand of the top 35 cities throughout the country is expected to double to 12,906 million cubic metres in 2021 as the population of these cities shoots up to 202 million from the present level of 107 million.

■ A handout released by the Human Settlement Management Institute, New Delhi suggested that even by the year 2001 we would have required an astounding 1.75 lakh crore for provision of all water infrastructures for urban domestic consumption.

If you still think these are not your concerns and definitely not your cup of tea, here's an excerpt from a recent news item in one of our own local dailies: "According to a senior citizen of Kumarpara, the problem has made life difficult for people immediately after the Durga Puja. Now, they are compelled to buy water and even that is not always possible.

"The problem faced by some other citizens of Santipur is also no less severe. For the last three days some residents are receiving filthy water through their taps. "The water is brownish and contains a foul smell," said Dr P Deka of Ashram Road, who added that it was not for the first time it has happened."

No matter how hard you try to escape your worries after you fill your buckets of water for the day, the fact is, anxiety always looms large on your mind for the next day's supply, especially if you reside in hilly areas of Jalla, Hengrabari, Geetanagar, Bhangagarh, etc, to name a few localities in Guwahati, and the situation is worse in winter. I do not really know whether to go by the statistics as I mentioned at the beginning, but I do have serious reservations and beg to differ on data provided by *Business Today*.

As a matter of fact, the picture is grimmer than what is meant to be. If the data released by the Ministry of Water Resources are to be believed, at 1250 cubic metres of water per person per year, we are, already a water-stressed nation. This has also been noticed in information released by Gardner-Outlaw & Engelman in 1997.

It is now estimated that by 1995 0.45 billion people in 31 countries experienced water-stress. By the year 2025, 48 countries with 2.8 billion people would experience water-stress condition and almost 4 billion people are expected to be living in water-stress condition in 54 countries worldwide by the year 2050. Although water seems to be a common commodity, of nearly 1370 million cubic kilometers of earth's total water quantity, only about two per cent of it is fresh out of which again nearly 97 per cent is tied up in polar caps in the form of permanent snow. The remaining three per cent is renewable through cycle of precipitation and evaporation from oceans and the surface sources on earth. The renewable fresh water falling on the continents and islands is only about 410,000 cubic kilometers, which is hardly 0.03 per cent of total quantity of water on earth.

Just six countries — Brazil, Russia, Canada, Indonesia, China and Colombo — account for half of world's fresh water supply. Iceland, with nearly 600,000 cubic metres and Guyana with nearly 315,000 cubic metres of water per inhabitant, rank near the top in terms of per capita availability of renewable fresh water and with 169 cubic meters and 55 cubic meters per capita, Jordan and United Arab Emirates rank near the bottom respectively. Not only the resources distributed unevenly among the countries, they are uneven within the boundaries of countries as well. Consider China; the country has about seven per cent of world's fresh water resources but most of this is concentrated on its southern parts.

Same is true for India as well. Even though the country has four per cent of world's fresh water resources, most of it lies along Brahmaputra basin covering Assam and adjoining areas with a per capita availability of 14,000 cubic meters. Sabarmati basin covering Gujarat region has less than 500 cubic meters. While parts of Meghalaya receive more than ten metres of rainfall during the year, arid and semi-arid regions of Rajasthan receive less than ten centimetres. The renewable resources are also not distributed equally in time. Even though New Delhi receives almost the same amount of rainfall as New York, the precipitation is evenly distributed throughout the year in New York, while the same amount falling within a span of about 100 days, that too within 100 hours, makes Delhi vulnerable to water scarcity.

Though the resources are not distributed evenly in time and space, in many parts of the world, large gaps in coverage has nothing to do with water scarcity. In fact, for universal access to 50 litres of water per person by 2015 we would require only one per cent of global water withdrawals. The world has enough water, but so far the political will and financial commitments to provide the poor with access to it is missing.

To meet the entire world's needs for clean water and sanitation it would cost an additional $9 billion per year well under $12 billion spent per year on perfumes and $18 billion spent on pet foods by Europe and North America alone. Today, one out of five people in developing world, i.e. 1.1 billion people, face daily risk of disease and death due to lack of access to clean drinking water. More than twelve million people faced famine in Ethiopia in 2003; yet 84 per cent of the flow of river Nile originates in Ethiopia's boundary. The country has no resources to harness its potential, whereas people in some water-rich countries are exploiting water as though it were a god-sent right.

Not only an average US citizen consumes 262 liters of water per day, the average US household, by consuming 10,000 KWH of electricity per year, is also indirectly consuming an additional 83,000 liters of water. A resident of Orange County, Florida, USA was billed for (can you believe!) 15.9 million litres of water for a single year. This is roughly equivalent to what 900 Kenyans use in a year. The disparity in allocation of resources lies within the boundaries of a country too. Consider the greater Guwahati water supply scheme. If sources at the Central Public Health and Environmental Engineering Organization are to be believed, the project costing Rs 399.48 crore, ranks lower in the priority list of the Urban Development Ministry, Government of India than at least two water supply projects from two other States costing much higher, even though the scarcity is conspicuous in this part of the region and the water resources ample.

Guwahati was declared a town in the 1853 and in the same year the town committee was formed. In 1865, Guwahati became a municipal town. Information available from learned sources shows that the last ever water treatment plant here was constructed by the State PWD at Satpukhuri in 1887. The scheme costing Rs 28,000 during those days covered an area of 4.5 square kilometres and a population of around four thousand. As per the official records, the population of the city was only 11,661 in the year 1901, which reached 84,601 in 1951 and 808,021 in the year 2001. Though the population of this city grew leaps and bounds during the last 50 years, there has not been any significant improvement in treated water supply facilities, especially since early 1980s.

A simple glance at a demand and supply pattern for the city will tell you where we stand today. Even if all the existing schemes (GMC — 71.50 MLD, PHED — 11.25 MLD, Assam Urban Water Supply & Sewerage Board — 12.60 MLD) in the city are considered to be running at their installed capacities, we would have a shortfall of more than 70 MLD by this time of the year. However, the treatment plants under Guwahati Municipal Corporation have already outlived their design period and are more than 40 years old and said to be running at less than half the capacities. Commissioned in 1977, the PHED treatment plant has already crossed 25 years of life span.

Therefore, it does not need a great deal of mathematics to safely calculate and conclude that present actual shortfall of drinking water supply in Guwahati city is no less than a staggering hundred million liters per day. Considering 15 per cent surcharge for unaccounted for water and a water demand of 135 litres per capita per day, this much quantity can cover a population of more than six lakh. This means, if no new schemes are taken up in near future, more than half the city population will live without piped water supply at any given point of time in future.

Let's now confabulate on what we can really do, till someone comes with that magic wand to ward off our water woes. The legendary singer Gordon Summer, popularly known as Sting of the famous pop band The Police sang, "When the world is running down, you make the best of what's still around." Though the song containing these lines might have been written with a different viewpoint, the entire philosophy of water conservation now should be to make the best of what we still have with us."

Here are some suggestions put forward by experts in the field.

■ Check all your home appliances using water for any leaks in the system. A leaky faucet can ruin 100-150 litres in a day.

■ Some old flushes drain an exorbitant amount of water each time they are used. It is always wise to use start and stop types.

■ A ten-minute shower will waste as much as 250 litres of water while a good bath needs no more than 55-75 litres.

■ Do not keep the tap running while brushing or washing your hands. 10-20 litres of water are wasted if you do so.

■ Select your model of washing machine that economises on water and power.

■ Adopt right landscape design to utilise water efficiently. Select plants that adapts to local climate more easily and do not require extra watering.

As a community, efforts in this direction could be to install individual rainwater harvesting systems in our homes. Experts predict that 15 per cent of all new building in Europe and the US will have an additional non-mains water source in 2010, mostly coming from rainwater collection.

We can also switch over to drip/sprinkler and other micro-irrigation systems to increase crop yields per litre of water consumed. Agriculture uses about 70 per cent of all the water extracted from earth's rivers, lakes and underground aquifers, and as much as 90 per cent in many developing countries.

We should also reduce losses in water distribution networks. Singapore reduced its unaccounted for water from 10.6 to 6.2 per cent from 1989 to 1995 and saved over $26 million in avoided capital facility expansions.

Industries should be relocated as per availability of water resources and water needed for the unit. An article titled 'Water-soluble ideas' posted on the web by Ted Sickinger of *The Oregon Staff* says, "It takes as much as 10 gallons to make a single computer chip. Viewed another way, a single chip plant can suck down up to 50 million gallons of water a month."

Well is that amount not enough for supply to 55,000 homes in our country. Think it over. Meanwhile, I go and fill my buckets; municipal water is running full force today.

(The writer can be contacted at: deka_m04@rediffmail.com)

Water Woes.
Mridul Deka
(As published in "The Assam Tribune", Guwahati, dtd- 17-12-2005)

(Although water is found abundantly on planet Earth, a majority of its population do not have access to clean drinking water. Guwahati too, despite being situated on the banks of the mighty Brahmaputra, is no exception with a regular shortfall of water, writes Mridul Deka.)

Six out of ten bore wells dug in north Gujarat and Saurashtra-Kutch region yield no water even at a depth of 1200 ft. In Loej village of Gujarat, each of the 200 families has one-member suffering from kidney problem. In Rajasthan, ten towns get water once in three days, and 31 once in two days. Twelve towns get it from 200 km away. Sixty-six million people in 200 districts risk fluoride contamination and 15 million risk arsenic poisoning. *(India Today,* June 9, 2003) Can we be an economic powerhouse without water? That's the question the country should be asking after per capita availability of fresh water has fallen from 5177 cubic meters to1869 cubic metres in 50 years. That's perilously close to 1700 mark fixed by the United Nations, below which we will become a 'water-stressed' nation. (Business Today; June 6, 2003)

Alarming statistics? Add some more to the list.
- About 1.1 billion people across the globe lack access to clean drinking water.
- More than 200 towns in India, excluding the class-1 cities, get less than 70 lpcd (litres per capita per day) of clean water. Compare this to 262 lpcd, average water consumption by US Citizens.

Union Ministry of Water Resources estimates that water demand of the top 35 cities throughout the country is expected to double to 12,906 million cubic metres in 2021 as the population of these cities shoots up to 202 million from from the present level of 107 million.

A handout released by the Human Settlement Management Institute, New Delhi suggested that even by the year 2001 we would have required an astounding 1.75 lakh crore for provision of all water infrastructures for urban domestic consumption.

If you still think these are not your concerns and definitely not your cup of tea, here's an excerpt from a recent news item in one of our own local dailies: "According to a senior citizen of Kumarpara, the problem has made life difficult for people immediately after the Durga Puja. Now, 'they are compelled to buy water and even that is not always possible.

"The problem faced by some other citizens of Santipur is also no less severe. For the last three days some residents are receiving filthy water through their taps. "The water is brownish and contains a foul smell," said Dr. P Deka of Ashram Road, who added that it was not for the first time it has happened."

No matter how hard you try to escape your worries after you fill your buckets of water for the day, the fact is, anxiety always looms large on your mind for the next day's supply, especially if you reside in hilly areas of Jatia, Hengrabari, Geetanagar, Bhangagarh, etc, to name a few localities in Guwahati, and when the season is winter. I do not really know whether to go by the statistics as I mentioned at the beginning, but I do have serious reservations and beg to differ on data provided by *Business Today*.

As a matter of fact, the picture is grimmer than what is meant to be. If the data released by the Ministry of Water Resources are to be believed, at 1250 cubic metres of water per person per year, we are, already a water-stressed nation. This has also been noticed in information released by Gardner- Outlaw & Engelman in 1997.

It is now estimated that by 1995, 0.45 billion people in 31 countries experienced water- stress. By the year 2025, 48 countries with 2.8 billion people would experience water- stress condition and almost 4 billion people are expected to be living in water-stress condition in 54 countries worldwide by the year 2050. Although water seems to be a common commodity, of nearly 1370 million cubic kilometers of earth's total water quantity, only about two percent of it is fresh out of which again nearly 97 percent is tied up in polar caps in the form of permanent snow. The remaining three per cent is renewable through cycle of precipitation and evaporation from oceans and the surface sources on earth. The renewable fresh water falling on the continents and islands is only about 410,000 cubic kilometers, which is hardly 0.03 per cent of total quantity of water on earth.

Just six countries - Brazil, Russia, Canada, Indonesia, China and Colombo - account for half of world's fresh water supply. Iceland, with nearly 600,000 cubic metres and Guyana with nearly 315,000 cubic metres of water per inhabitant, rank near the top in terms of per capita availability of renewable fresh water and with 169 cubic metres and 55 cubic meters per capita. Jordan and United Arab Emirates rank near the bottom respectively. Not only the resources distributed unevenly among the countries they are uneven within the boundaries of countries as well. Consider China; the country has about seven per cent of world's fresh water resources but most of this is concentrated on its southern parts.

Same is true for India as well. Even though the country has four per cent of world's fresh water resources, most of it lies along Brahmaputra basin covering Assam and adjoining areas with a per capita availability of 14,000 cubic meters. Sabarmati basin covering Gujarat region has less than 500 cubic meters. While parts of Meghalaya receive more than ten metres of rainfall during the year, arid and semi-arid regions of Rajasthan receive less than ten centimetres. The renewable resources are also not distributed equally in time. Even though New Delhi receives almost the same amount of rainfall as New York, the precipitation is evenly distributed throughout the year in New York, while the same amount falling within a span of about 100 days, that too within 100 hours, makes Delhi vulnerable to water scarcity.

Though the resources are not distributed evenly in time and space, in many parts of the world, large gaps in coverage has nothing to do with water scarcity. In fact, for universal access to 50 litres of water per person by 2015 we would require only one per cent of global water withdrawals. The world has enough water, but so far, the political will and financial commitments to provide the poor with access to it is missing.

To meet the entire world's needs for clean water and sanitation it would cost an additional $9 billion per year, well under $12 billion spent per year on perfumes and $18 billion spent on pet foods by Europe and North America alone. Today, one out of five people in developing world, i.e. 1.1 billion people, face daily risk of disease and death due to lack of access to clean drinking water. More than twelve million people faced famine in Ethiopia in 2003; yet 84 per cent of the flow of river Nile originates in Ethiopia's boundary. The country has no resources to harness its potential, whereas people in some water-rich countries are exploiting water as though it were a god-sent right.

Not only an average US citizen consumes 262 liters of water per day, the average US household, by consuming 10,000 KWH of electricity per year, is also indirectly consuming an additional 83,000 liters of water. A resident of Orange County, Florida, USA was billed for (can you believe!) 15.9 million litres of water for a single year. This is roughly equivalent to what 900 Kenyans use in a year. The disparity in allocation of resources lies within the boundaries of a country too. Consider, the greater Guwahati water supply scheme; if sources at the Central Public Health and Environmental Engineering Organization are to be believed, the project costing Rs. 399.48 crore, ranks lower in the priority list of the Urban Development Ministry, Government of India than at least two water supply projects from two other States costing much higher, even though the scarcity is cornspacious in this part of the region and the water resources ample.

Guwahati was declared a town in the 1853 and in the same year the town committee was formed. In 1865, Guwahati became a municipal town. Information available from learned sources shows that the first ever water treatment plant here was constructed by the State PWD at Satpukhuri in 1887. The scheme costing Rs. 28,000 during those days covered an area of 4.5 square kilometres and a population of around four thousand. As per the official records, the population of the city was only 11,661 in the year 1901, which reached 84,601 in 1951 and 808,021 in the year 2001. Though the population of this city grew leaps and bounds during the last 50 years, there has not been any significant improvement in treated water supply facilities, especially since early 1980s.

A simple glance at a demand and supply pattern for the city will tell you where we stand today. Even if all the existing schemes, (GMC-71.50MLD, PHED-11.25MLD, Assam Urban Water Supply & Sewerage Board-12.60MLD) in the city are considered to be running at their installed capacities, we would have a shortfall of more than 70 MLD by this time of the year. However, the treatment plants under Guwahati Municipal Corporation have already outlived their design period and are more than 40 years old and said to be running at less than half the capacities. Commissioned in 1977, the PHED treatment plant has already crossed 25 years of life span.

Therefore, it does not need a great deal of mathematics to safely calculate and conclude that present actual shortfall of drinking water supply in Guwahati city is no less than a staggering hundred million liters per day. Considering 15 per cent surcharge for unaccounted for water and a water demand of 135 liters per capita per day, this much quantity can cover a population of more than six lakh. This means, if no new schemes are taken up in near future, more than half the city population will live without piped water supply at any given point of time in future.

Let us now confabulate on what we can really do, till someone comes with that magic wand to ward off our water woes-. The legendry singer Gordon Summer, popularly known as Sting of the famous pop band The Police sang, "When the world is running down, you make the best of what's still around." Though the song containing these lines might have been written with a different viewpoint, the entire philosophy of water conservation now should be to make the best of what we still have with us.

Here are some suggestions put forwarded by experts in the field.

Check all your home appliances using water 'for any leaks in the system. A leaky faucet can ruin 100-150 litres in a day.

Some old flushes drain an exorbitant amount of water each time they are used. It is always wise to use start and stop types.

A ten-minute shower will waste as much as 250 litres of water, while a good bath needs no more than 55-75 litres.

Do not keep the tap running while brushing or washing your hands, 10-20 litres of water are wasted if you do so.

Select your model of washing machine that economizes on water and power.

Adopt right landscape design to utilize water efficiently. Select plants that adapts to local climate more easily and do not require extra watering.

As a community, efforts in this direction could be to install individual rainwater harvesting systems in our homes: Experts predict that 15 per cent of all new building in Europe and the US will have an additional non-mains water source in 2010, mostly coming from rainwater collection.

We can also switch over to drip/sprinkler and other micro irrigation systems to increase crop yields per litre of water consumed. Agriculture uses about 70 per cent of all the water extracted from earth's rivers, lakes and underground aquifers, and as much as 90 percent in many developing countries.

We should also reduce losses in water distribution networks: Singapore reduced its unaccounted-for water from 10.6 to 6.2 per cent from 1989 to 1995 and saved over $26 million in avoided capital facility expansions.

Industries should be relocated as per availability of water resources and water needed for the unit. An article titled 'Water-soluble ideas' posted on the web by Ted Sickinger of *The Oregon Staff* says, "It takes as much as 10 gallons to make a single computer chip. Viewed another way, a single chip plant can suck down up to 50 million gallons of water a month."

Well is that amount not enough for supply to 55,000 homes in our country. Think it over. Meanwhile, I go and fill my buckets; municipal water is running full force today.

(The writer can be contacted at: deka_m04@rediffmail.com)

The Assam Tribune, Guwahati, Saturday, December 13, 2003 — horizon — Page III

Harvesting the rain

Down the STREAM
Ron Dwarah

A national shame

Agreed that New Delhi does not treat Assam well, and railway ministers care little for the unemployed of the North-east. What has taken place in the past few weeks in Assam has paid red-green down well with thousands of well-meaning Assamese. It just should not be that if one person of the North-east is beaten up by anti-socials in Bihar, the retaliation should be in a tit for tat basis. Let us look at the whole issue dispassionately. There are unemployed youths in all the States of the North-east, just as there are unemployed youths in Bihar. It cannot be that Bihari youth's fault that they came to Guwahati to take a test for some railway em-

ployment examination. But it is certainly the fault of hoodlums in Assam that these visitors were roughed up on the streets — the photo of a Bihari youth being kicked about on the road in a busy Guwahati locality by Assamese goondas has immensely damaged the Assamese image — which led to all the complications. This was followed by hooligans in Bihar taking over several railway stations there and harassing, molesting and torturing railway passengers from the North-east. The incidents in Bihar resulted in widespread arson in Assam, with Bihari homes and shops being set ablaze, their property looted and as many as 56 Biharis losing their lives.

Unlike the other settlers in Assam, most of the Biharis living in the State have almost fully assimilated with the mainstream Assamese society. As they mostly live in the interior places, many have even lost contact with their relatives back in Bihar. And unlike the other trading communities that have made their millions through questionable means in Assam, the Biharis accumulated their wealth by sheer dint of hard work. The Biharis have in thousands of instances have associated themselves with activities of the Assamese society unlike the other trading communities, which have avoided the joys and tragedies of the native society. It is under such circumstances that journalists visiting the relief camps of Bihari refugees in the Dibrugarh and Tinsukia districts left to describe. As the refugees spoke about their horror tales in chaste Assamese, one could not but empathise.

It is under such circumstances that the need of the hour is for the Assamese to make the first move in rehabilitating the estranged Biharis of Assam. We certainly do not want a Laloo Prasad Yadav to tell us what to do! The native community

L et us first go through a few facts and figures about the status of water around the world in general and in India in particular. About 1.1 billion people across the globe lack access to clean drinking water and two billion people around the world depend on ground water for their survival. Eighty countries around the world experienced water shortage by mid 1990. This is roughly equivalent to 40 per cent of world population. About two-third of the global population is expected to be living in water-stressed condition in less than 25 years. An amount of $30 billion per year will be needed for universal access to water by the year 2015. These are some of the figures released by the United Nations Environment Program (UNEP) last year.

In India, 90 per cent of water consumption is through agriculture, six per cent of water consumption is through the industries and only four per cent of water available are utilised for domestic consumption. The UN has set 1700 cubic meters as the per capita availability of fresh water below which the country would be a 'water-stressed' nation. The present per capita availability of water is 1869 cubic meters, which was 5177 cubic meters 50 years before. More than 200 towns excluding the class-I cities, get less than 100 lpcd (litres per capita per day) of water.

Compare this with the supply level of some advanced countries — it is 350 lpcd. The Union Ministry of Water Resources estimates that water demand of the top 35 cities throughout the country is expected to double by 2021 as the population of these cities shoots up to 202 million from the present level of 107 million. A handout released by the Human Settlement Management Institute, New Delhi suggested that even by the year 2001 we would have required an astounding Rs 1.75 lakh crore for provision of all water infrastructures for urban domestic consumption.

A pertinent question here is where do we get all this money. But the grim reality is that even if we managed to have such a huge amount to meet the demand for water infrastructure, perhaps there is not enough supply to meet the demand. About 85 per cent of the rural population and 55 per cent of the urban population in India depend on ground water. Delhi Jal Board depends on 1250 deep tube-wells for ground water in addition to its surface sources. As per a World Bank assessment, about 30 per cent of water output is lost that is nearly 800 mld of water finds its way through the city drains in simple words, all the 1250 deep tube-wells work round the clock and 800 million litres of water daily through the city gutters. The country has some 56 lakh bore wells, which have lowered the ground water to a critical level. In 206 out of 597 districts across the country. Long term

dependence on ground water without replenishment/recharging of sources through water harvesting and concurrent rise in population has already had a telling effect on the nation's economy and per capita availability of fresh water. No serious attempts have been made yet to harvest rainwater at macro or micro level. Monsoon accounts for 80 per cent of rainwater. During the monsoon season, i.e. June to September, the country receives on an average 88 cm of rainfall. Monsoon is considered normal when the rainfall associated with it, is +/-10 per cent of its long term average. The effect of rainfall associated with monsoon is overwhelming on India's economy. The worst case scenario was in the year 2010, when Punjab reported a low of nearly Rs 800 crore due to 49 per cent rainfall deficit. Andhra Pradesh reported a loss of Rs 610 crore due 39 per cent deficit and Rajasthan re

ported a loss of Rs 4996 crore due to a rainfall deficit of 67 per cent.

The country has about four per cent of world's fresh water resources. The total rainwater potential of the country is about 2500 billion cubic meters. This translates into roughly 1869 billion cubic meters of river water (approximately twice the size of river Brahmaputra, of which, only 690 billion cubic meters is only utilised. The rest of it flows back to sea. Due to insufficient storage, we face water shortage and droughts. With a good rainwater management system the water problem of the country may be reduced if not solved.

Basically rainwater harvesting systems are of two types:
1) Rainwater harvesting systems for domestic consumption. Here the rainwater is collected and stored in a storage tank during the rain and utilised for drinking, cooking, etc. later in scarcity during the dry period.
2) Rainwater harvesting systems for ground water recharge. Here the rainwater is collected during the rain and passed on to recharge trenches for ground water recharge or the rainwater is passed on to an aquifer below the ground through tube wells made for the purpose.

There are basically seven components for rainwater harvesting system. They are

Catchment: It is the surface, which directly receives rainwater and provides water to the system. It may be a paved surface like terrace of a building or the courtyard of the building or it may be an unpaved surface within the compound like a lawn or open ground. Thus an RCC roof or a CGI sheet roof also acts as a catchment of a rainwater harvesting system.

Coarse mesh: It is a mesh of screening materials to prevent debris from entering the rainwater pipes. It is placed at the inlet to the pipes.

Gutters: They are the channels provided all around the roof surface to collect rainwater. The size of the gutters should be able to take discharge from the rainfall of highest intensity.

Conduits: These are the pipelines that carry rainwater from the rooftop to storage tanks or to the recharge pit. These pipes may be PVC or GI depending upon the choice of the user. In general, PVC pipes are preferred due to their lower costs and easy handling.

First flushing: The first flushing device is specially used to drain off the dirty rainwater from the first shower so that it does not enter the system. This may be a Tee with one end plugged or any other suitable device to drain out the first shower of rainwater.

Filter: After the first flush of rain is passed out it is allowed to pass through a filter. The various types of filters, presently being used are sand filter, charcoal filter, Dewas filter, Jeyakumar filter.

Storage tanks: The storage tanks or the cistern is one of the important components of a rainwater harvesting system. This is because the type and capacities of such tanks used for domestic consumption, affect investment decisions on such systems. These also include Ferrocement storage tanks.

The rapidly growing population in our cities and subsequent water demand, coupled with dependence on ground water sources, has not only reduced the ground water level of our supply sources but it has also put a question mark on future sources of water supply. Thus, solution to the present water scarcity perhaps lies partially in effectively harvesting the rainwater for domestic consumption and recharge of ground water sources. An effort to harvest rainwater at macro and micro level could reduce our water problems, if not mitigate it completely.

Mridul Deka

Harvesting the rain
Mridul deka
(As published in 'The Assam Tribune',
Dtd – 13/12/2003)

Let us go through few facts and figures about the status of water around the world in general and in India in particular. About 1.1 billion people across the globe lack access to clean drinking water and two billion people around the world depend on ground water for their survival. Eighty countries around the world experienced water storage by mid-1990. This is roughly equivalent to 40 per cent of world population. About two-third of the global population is expected to be living in water-stressed condition in less than 25 years. An amount of $30 billion per year will be needed for universal access to water by the year 2015. These are some of the figures released by the United Nations Environment Program (UNEP) last year.

In India, 90 per cent of water consumption is through agriculture, six per cent of water consumption is through the industries and only four per cent of water available are utilized for domestic consumption. The UN has set 1700 cubic meters as the per capita availability of fresh water below which the country would be a 'water-stressed' nation. The present per capita availability of water is 1869 cubic meters, which was 5177 cubic meters 50 years before. More than 200 towns excluding the class-I cities, get less than 100 lpcd (liters per capita per day) of water.

Compare this with the supply level of some advanced countries – it's 350 lpcd. The Union Ministry of Water Resources estimates that water demand of the top 35 cities throughout the country is expected to double to 12,906 million cubic meters in 2021 as the population of these cities shoots up to 202 million from the present level of 107 million. A handout released by the Human Settlement Management Institute, New Delhi suggested that even by the year 2001 we would have required as astounding Rs1.75 lakh crore for provision of all water infrastructure for urban domestic consumption.

A pertinent question here is where do we get all this money. But the grim reality is that even if we managed to have such a huge amount to meet the demand for water infrastructure, perhaps there is not enough supply to meet the demand. About 85 per cent of the rural population and 55 per cent of the urban population in India depend on ground water. Delhi Jal Board depends on 1250 deep tube-wells for ground water in addition to its surface sources. As per a World Bank assessment, about 30 per cent of water output is lost that is nearly 800 mld of water finds their way through the city drains. In simple words, all the 1250 deep tube-wells work round the clock and send 800 million liters of water daily through the city gutters. The country has some 56 lakhs bore wells, which have lowered the ground water to a critical level in 206 out of 597 districts across the country. Long term dependence on ground water without replenishment/recharging of sources through water harvesting and concurrent rise in population has already had a telling effect on the nation's economy and per capita availability of fresh water. No serious attempts have been made yet to harvest rainwater at macro or micro level. Monsoon accounts for 80 per cent of rainwater. During the monsoon season, i.e. June to September, the country receives on an average 88 cm of rainfall. Monsoon is considered normal when the rainfall associated with it, is +/-10 per cent of its long-term average. The effect of rainfall associated with monsoon is overwhelming on India's economy. The worst-case scenario was in the year 2000, when Punjab reported a loss of nearly Rs. 800 crore due to 49 per cent rainfall deficit. Andhra Pradesh reported a loss of Rs. 610 crore due to 39 per cent deficit and Rajasthan reported a loss of Rs. 4996 crore due to a rainfall deficit of 67 per cent.

The country has about four per cent of world's fresh water resources. The total rainwater potential of the country is about 2500 billion cubic meters. This translates into roughly 1869 bcm (billion cubic meters) of river water (approximately twice the size of river Brahmaputra, of which, only 690 billion cubic meters is only utilized. The rest of it flows back to sea. Due to insufficient storage, we face water shortage and droughts. With a good rainwater management system, the water problem of the country may be reduced if not solved.

Basically, rainwater harvesting systems are of two types.
- Rainwater harvesting systems for domestic consumption: Here the rainwater is collected and stored in a storage tank during the rain

and utilized for drinking, cooking, etc. later in scarcity during the dry period.
- Rainwater harvesting systems for ground water recharge. Here the rainwater is collected during the rain, and passed on to recharge trenches for ground water recharge or the rainwater is passed on to an aquifer below the ground through tube wells made for the purpose.

There are basically seven components for harvesting rainwater. They are
1. *Catchment:* It is the surface, which directly receives rainwater and provides water to the system. It may be a paved surface like terrace of a building or the courtyard of the building or it may be an unpaved surface within the compound like a lawn or open ground. Thus, an RCC roof or a CGI sheet roof also acts as a catchment of a rainwater harvesting system.
2. *Coarse mesh:* It is a mesh of screening materials to prevent debris from entering the rainwater pipes. It is placed at the inlet to the pipes.
3. *Gutters:* They are the channels provided all-round the roof surface to collect rainwater. The size of the gutters should be able to take discharge from the rainfall of highest intensity.
4. *Conduits:* These are the pipelines that carry rainwater from a rooftop to storage tanks or to the recharge pit. These pipes may be PVC or GI depending upon the choice of the users. In general, PVC pipes are preferred due to their lower costs and easy handling.
5. *First flushing:* The first flushing device is specially used to drain off the dirty rainwater from the first shower so that it does not enter the system. This may be a Tee with one end plugged or any other suitable device to drain out the first shower of rainwater.
6. *Filter:* After the first flush of rain is passed out it is allowed to pass through a fibre. The various types of filters, presently being used are sand filter, charcoal filter, Dewas filter, Jeyakumar filter.
7. *Storage Tanks:* The storage tanks or the cistern is one of the important components of a rainwater harvesting system. This is because the type and capacities of such tanks used for domestic consumption affect investment decisions in such systems. These also include Ferrocement storage tanks.

The rapidly growing population in our cities and subsequent water demand, coupled with dependence on ground water sources, has not only reduced the ground water level of our supply sources but it has also put a question mark on future sources of water supply. The solution to the present water scarcity perhaps lies partially in effectively harvesting the rainwater for domestic consumption and recharge of ground water sources. An effort to harvest rainwater at macro and micro level could reduce our water problems, if not mitigate it completely.

horizon
Saturday Special

The Assam Tribune, Saturday, July 3, 1999

A priceless PROPOSITION

The rapid urbanisation in the country has led to scarcity of basic amenities, drinking water being one of them. The country today has more than 25 'million-plus' cities, Guwahati included, but majority of the population in these cities do not have access to safe drinking water. **Mridul Deka** looks into the country's water scenario.

As the rolling tanks of barrashed sand merge into the muddy golden hues of twilight Anca, 50, crouches over the piece of water bubbling up from the dune. She quickly fills her earthen pot and replaces the plug. It's a routine she follows once a week, the only time water courses through the pipe, as it heads for its official outlet at a village 8 km away. To Anca, a simple caretaker in this parched corner of southwestern Rajasthan, the pipeline is a god send. "Life is easier now", She smiles. No longer is she dead tired, as she used to be, hauling water 200 lt from the deepening depth of the village well. Still, every person in the 12 houses in Anca's tiny hamlet of Neera in Dhara Bharatiad, ironically, as "place of monsoon." In Churu district gets by with just about 5 litres of water a day, less than the water that flows out of a tap when you brush your teeth.

That was an excerpt from an article that appeared in a national weekly some time back in 1998. And if you thought that, that was the situation in the dry, arid of Rajasthan, do not be fooled by your complacency. Mr Ishwar Chand Bajaj, a resident of Lanka town in Nagaon district and a founder member of the Lanka Chambers of Commerce has this to say: "I have never seen such scarcity of water throughout my life. The old water supply scheme of PHE Department is more than 40 years old and has become irrelevant at present. It serves only a handful of population of our town and only last week due to a fault in their distribution system, we were left with no piped water for almost five days and those who have the most disadvantageous would not let others use there as water level has receded at least 30 ft to 15 ft this year due to drought-like situation. And there, one fine morning in Delhi roads, I was surprised to see a not-so-small crowd gathering on the road just behind the police station. Getting near, I could not have been more amused as myself as to how one was loaded by one a preconceived notion of a mob situation only to find, here was, a gathering of people from all ages, men, women, children, thronging to collect their share of water from this damaged stand-ie, which leaks every time the water passes through it at around 9-10 in the morning.

This is one face of rapid urbanisation in the country, and the corresponding scarcity of basic amenities. The urban population itself has increased four-fold from 62 million in 1951 to 217 million in 1991. On a percentage basis, of the total population in the country, 17.3 per cent of total population lived in urban India in 1951 which has increased to 25.7 per cent till 1991, and it experts are to be believed, this mix of urbanisation would accelerate more once 25 per cent limit has been crossed. However in case of Assam, the share of urban population to the total of total population is 11.1 per cent which is below the national average. The country had only five cities where population exceeded one million in 1951, this has grown to twenty-three in 1991 and at present there are around twenty-seven 'million-plus' cities in the country, 'Guwahati' being one of them.

Statistics will tell you, as of 1991, 81.38 per cent of urban households covering 88 per cent of the urban population has access to safe drinking water. Statistics lie. The per capita consumption of water for the country as a whole at 71 litres per capita per day (lpcd) is lower than, the prescribed norms.

Indian Standard 1172, Code of Basic Requirements of Water Supply, Drainage and Sanitation and National Building Code, recommend a minimum of 135 lpcd for all residences provided with full flushing systems for excreta disposal. The scope of IS 1172 states that it does not take into account overcrowded/both casual and permanent and is based on the assumption that average size of a family is five and 9 sq m floor area is occupied by not more than two persons.

The Central Public Health and Environmental Engineering Organisation (CPHEEO), Government of India, considers the following rates per capita per day minimum for domestic and non-domestic needs as a general rule.

■ For communities with population up to 20,000
(i) Water supply through stand posts (min) = 40 pcd
(ii) Water supply through house connections = 70 to 100 lpcd
■ For communities with population from 20,000 to 100,000–3000 to 150 lpcd
■ For communities with population above 100,000–150 to 200 lpcd

The norms suggested by the working group on urban water supply and sanitation sector for Ninth Five Year Plan (1997-2002) are as follows:

■ 125 lpcd for urban areas, where piped water supply and underground sewerage system are available
■ 70 lpcd for urban areas, provided with piped water supply but without underground sewerage system
■ 40 lpcd for towns with spot sources/stand posts. One source for 20 families within walking distance of 100 metres.

Reports on Seventh Five Year Plan also suggest that twenty-one 'lakh of urban areas did not have access to safe drinking water supply. Though the situation is much more critical than this on field. Even at the present rate of daily supply, the country will need additional 6,500 million cu/m of water supply by the year 2001. The cost, believe it or not, is simply Rs 1.95 lakh crore.

And for Guwahati with a total population of around 10 lakh, the total installed capacity of all water treatment plants located at Panbazar, Satpukhuri, Guwahati University, Kamakhya Hills, Panda, Guwahati Refinery and Hengerabari is 1.14 (62) mld (million litres per day). However except in the treatment plant at Hengerabari, constructed by the Assam Urban Water Supply and Sewerage Board with financial assistance from Housing & Urban Development Corporation (HUDCO), all plants are working much below their capacities. Even it we consider these plants to be working at 80 per cent of their installed capacities, the 10 mgd water treatment plant at Panbazar is said to be working at 50 per cent of its capacity, the daily supply of water for the city is 91,889 mld. This does not account for the unwanted losses in distribution system. As per findings of pilot projects in selected cities in India on the ratio of leakage in distribution system carried out by the National Environmental Engineering Research Institute (NEERI), Nagpur, the leakage losses in water mains, service connections, wastes were found to be 20 to 36 per cent of the total flow in the system, which is categorised as excessive by the Manual of Water Supply and water from the ground water table are all tucked away. Its wastage ladden borewells meant her irrigation. At community well at Bharapura, at data in Guwahati say, about 25 years ago to cater to the need of some 30 families is now left with one, 5-6 ft in water, approximately 1.6 cum catering to more than 20 families at present. Similar is the situation to other localities of the city as well.

Though, the National Water Policy states that domestic drinking water has the highest priority in terms of water allocation, at present there are no laws or legislations framework which governs supply, connection, every missing link.

When Prof Sanraj Surma from the Human Settlement Management Institute, New Delhi, during a WHO-sponsored training programme, on urban community water supply and sanitation in Guwahati raised the horrors committing powers/house construction to Guwahati as either resettlement or urban local bodies, as to what would however, if the state of water he was leading, apposedly of his heirs, the answer was obvious — the place would break. But the professor himself disappointed, it, to make everyone realise how bigger have — the water should gold. And how precious is every hector of land, all city dwellers are beginning to realise block, be it the small town like Lanka in Nagaon district or in a large urbanized million-plus city like Guwahati.

So, next time we just a honey is hit, to be too much consumption, do not hurt over it. Its can be aptly. Because the emotions behind, for a teacher, in the remote of the town, he mainly, whit, yours it goes through, will convoying his own word have to one understand the your product, hill carring the way to terms of water being the losses emanating from an entire hunger in their farmers, dog set, behind you water mouths as the city the goes down as water.

A priceless Proposition

Mridul deka
(As published in 'The Assam Tribune',
Dtd – 03/07/1999)

(The rapid urbanization in the country has led to the scarcity of basic amenities, drinking water being one of them. The country today has more than 25 million-plus' cities, Guwahati included, but majority of the population in these cities do not have access to safe drinking water. Mridul Deka looks into the country's water scenario.)

"As the rolling banks of burnished sand merge into the muddy golden haze of twilight, Anni, 50, crouches over the pool of water bubbling up from the dune. She quickly fills her earthen pot and replaces the plug. It's a routine she follows once a week, the only time water courses through the pipe, as it heads for its official outlet at a village 8 km away. To Anni, a temple caretaker in this parched corner of north-western Rajasthan, the pipeline is a god sent "Life is easier now". She smiles. No longer is she dead tired as she used to be, hauling water 200 ft from the deepening depth of the village well. Still, every person in the 12 houses in Anni's tiny hamlet of Nami ki Dhani (translated, ironically, as "place of moisture") in Churu district gets by with just about 5 litres of water a day, less than the water that flows out of a tap when you brush your teeth".

That was an excerpt from an article that appeared in a national weekly some time back in 1998. And if you thought that, that was the situation in the dry land of Rajasthan, do not be fooled by your complacency. Mr. Inder Chand Bajaj, a resident of Lanka town in Nagaon district and a founder member of the Lanka Chambers of Commerce, has this to say.

"I have never seen such scarcity of water throughout my life. The old water supply scheme of PHE Department is more than 40 years old and has become irrelevant at present. It serves only a handful of population of our town and only last week due to a fault in their distribution system, we were left with no piped water for almost five days and those who have their individual sources would not let others use them as water level has receded at least 10 ft to 15 ft in this year due to drought-like situation." And there, one fine morning of Diphu town, I was surprised to see a not-so-small crowd gathering on the road just behind the police station. Getting near, I could not have been more amused at myself as to how one was fooled by one's pre-conceived notion of a mob-skirmish only to find, here was, a gathering of people from all ages, men, women, children, thronging to collect their share of water from this damaged air-valve, which leaks every time the water passes through it at around 9-10 in the morning.

This is one face of rapid urbanization in the country and the corresponding scarcity of basic amenities. The urban population itself has increased four-fold from 62 million in 1951 to 217 million in 1991. On a percentage basis, of the total population in the country, 17.3 per cent of total population lived in urban India in 1951 which has increased to 25.7 per cent till 1991, and if experts are to be believed, this rate of urbanization would accelerate faster once 25 per cent limit has been crossed. However, in case of Assam, the share of urban population as percentage of total population is 18.8 per cent which is below the national average. The country had only five cities where population exceeded one million in 1951, this has grown to twenty-three in 1991 and at present there are around twenty-seven million-plus cities in the country, Guwahati being one of them.

Statistics will tell you, as of 1991, 81.38 per cent of urban households covering 85 per cent of the urban population have access to safe drinking water. Statistics lie. The per capita consumption of water for the country as a whole at 71 litres per capita per day (lcpd) is lower than the prescribed norms.

Indian Standard 1172, Code of Basic Requirements of Water Supply, Drainage and Sanitation and 'National Building Code', recommend a minimum of 135 lpcd for all residences provided with full flushing systems for excreta disposal. The scope of IS 1172 states that it does not take into account overcrowding both casual and permanent and is based on the assumption that average size of a family is five and 9 sq mtr floor area is occupied by not more than two persons.

The Central Public Health and Environmental Engineering Organisation (CPHEEO), Government of India, considers the following rates per capita per day minimum for domestic and non-domestic needs, as a general rule:

For communities with population up to 20,000
(i) Water supply through stand posts (min) = 40 lpcd
(ii) Water supply through house connections = 70 to 100 lpcd.

For communities with population from 20,000 to 100,000, = 100 to 150 lpcd.
For communities with population above 100,000, =150 to 200 lpcd.

The norms suggested by the working group on urban water supply and sanitation sector for Ninth Five Year Plan (1997-2002) are as follows:
- 125 lpcd for urban areas provided with piped water supply and underground sewerage system are available.
- 70 lpcd for urban areas provided with piped water supply but without underground sewerage system.
- 40 lpcd for towns with spot sources/stand posts.

One source for 20 families within walking distance of 100 meters.

Reports on Seventh Five Year Plan also suggest that nearly one third of urban areas did not have access to safe drinking water supply. Though, the situation is much more critical than this on field. Even at the present rate of daily supply, the country will need additional 6,500 million cum of water supply by the year 2001. The cost, believe it or not, is simply Rs. 1.95 lakh crore.

And for Guwahati with a total population of around 10 lakh, the total installed capacity of all water treatment plants located at Panbazar, Satpukhuri, Guwahati University, Kamakhya Hills, Pandu, Guwahati Refinery and Hengrabari is 114.862 Mld (million litres per day). However, except for the treatment plant at Hengranbari, constructed by the Assam Urban Water Supply and Sewerage Board with financial assistance from Housing & Urban Development Corporation (HUDCO), all plants are working much below their capacities. Even if we consider these plants to be working at 80 per cent of their installed capacities (the 10 mgd water treatment plant at Panbazar is said to be working at 50 per cent of its capacity), the daily supply of water for the city is 91,889 mld. This does not account for the onward losses in distribution system. As per findings of pilot projects in selected cities in India on the status of leakage in distribution system carried out by the National Environment Engineering Research Institute (NEERI), Nagpur, the leakage losses in water mains, service connection, valves were found to be 20 to 36 per cent of the total flow in the system, which is categorised as excessive by the Manual of Water Supply and Treatment of the CPHEEO (with up to 10 per cent as low, 10-20 per cent as average, 20-50 per cent as excessive and above 50 percent as alarming). If losses in distribution system are put at 25 per cent for Guwahati of the 91.889 mld, only 68.916 mld reaches the city population. And at standard norms this caters to a population of around 510,500. This leaves almost half the city's population with no access to piped water supply. With half a million people depending on their individual ground water sources for their daily usage of water, the situation has already reached a critical stage.

Nearly 85 per cent of rural drinking water and 55 per cent of urban drinking water comes from ground water sources. But most water from the ground water basins are sucked away by unregulated borewells meant for irrigation. A community well at Bhogargaon at Jatia in Guwahati dug about 15 years ago to cater to the needs of some 10 families is now left with only 5-6 ft of water, approximately 1.4 cum, catering to more than 20 families at present. Similar is the situation in other localities of the city as well.

Though the National Water Policy states that, domestic drinking water has the highest priority in terms of water allocation, at present there are no legal or administrative framework which allows for reallocation of water from low value agricultural use to high value urban and industrial use. More than 85 per cent of all of India's water is used for irrigation. The country today is littered with nearly 300 unfinished irrigation projects. Of the 119 major dams and canals, 24 have dragged on for more than 25 years. (Saryu Canal in Uttar Pradesh started in 1976 is incomplete even today). The money needed to finish these projects is Rs. 41,272 crores. But the real cost after all these spendings, less than five per cent of the country's water is available for drinking purpose. The scarcity and concomitant anxiety become more conspicuous in form of petitions and prayers reaching local water supply authorities for water supply connections every passing day.

When Prof. Sanjay Sarma from the Human Settlement Management Institute, New Delhi, during a WHO sponsored training programme on urban community water supply and sanitation at Guwahati asked the trainees (comprising engineers, doctors, bureaucrats and elected representatives of urban and local bodies), as to what would happen if the glass of water he was holding slipped out of his hand, the answer was obvious – the glass would break. But the professor himself disapproved, only to make everyone realize what he believed was more important – the water would spill. And how precious every bucket of water is, all city-dwellers are beginning to realize slowly, be it in a small town like Lanka in Nagaon district or in a more urbanized million-plus city like Guwahati.

So next time you get a fat water bill for too much consumption, do not fret over it, but use it rationally. Because the amount you pay for a bucket of treated water may conveniently justify what you would have to pay otherwise for your medical bill coming on your way in terms of water-borne disease emanating from an untreated source or a hefty capital investment on your privately dug well behind your campus only to find it go dry after 4-5 years.

www.ingramcontent.com/pod-product-compliance
Lightning Source LLC
Chambersburg PA
CBHW030926180526
45163CB00002B/475